Richard M. Hodges

Practical Dissections

Richard M. Hodges
Practical Dissections
ISBN/EAN: 9783337250164
Printed in Europe, USA, Canada, Australia, Japan
Cover: Foto ©berggeist007 / pixelio.de

More available books at **www.hansebooks.com**

BY

RICHARD M. HODGES, M.D.,

FORMERLY DEMONSTRATOR OF ANATOMY IN THE MEDICAL DEPARTMENT OF HARVARD UNIVERSITY.

SECOND EDITION, THOROUGHLY REVISED.

PHILADELPHIA:
HENRY C. LEA.
1867.

Entered according to Act of Congress, in the year
R. M. HODGES,
in the Clerk's Office of the District Court of the District

PHILADELPHIA:
COLLINS, PRINTER, 705 JAYNE STREET.

NOTE TO THE SECOND EDITION.

In revising the following pages no alterations have been made, other than those which experience in their use has suggested. It has been the Author's endeavor to make the descriptions as clear and concise as possible, rather than to add to their details; and to render the volume, in all respects, more deserving of the favor it has received.

The few pages on Anatomical Landmarks were suggested by, and to a small extent taken from, an article by Mr. Luther Holden, contained in the second volume of the "Reports of St. Bartholomew's Hospital."

BOSTON, *February,* 1867.

PREFACE.

The "Practical Dissections" is not a Treatise on Anatomy, nor in any way a substitute for one. It is intended to be simply a practical guide in the ordinary dissections of the Medical Student, describing on the same page, and in connection, the muscles, nerves, arteries, veins, or other structures which are conjointly exposed, and only so far as exposed, in dissecting any one of the parts into which the dead subject is usually divided. Remembering that "the smallness of the size of a book is always its own recommendation, as, on the contrary, the largeness of a book is its own disadvantage, as well as the terror of learning," all Minute Anatomy, and the details of arterial distribution, beyond what an ordinary injection exhibits, or of nervous ramifications which only special dissections can demonstrate, have purposely been omitted, or, if introduced, as has been done almost of necessity in a few places, are accompanied by the statement that their verification in an ordinary dissection is not to be expected.

The order observed in the following pages is an entirely arbitrary one, but which an experience of seven years in demonstrating, and more than ten in the special study of Anatomy, has shown to be the most convenient in the Dissecting Room, and the most economical of material. The division of the descriptions into "dissections," each intended to comprise a day's work; it is believed will be found advantageous, not only as mapping out the labor before the Dissector, but in giving him an opportunity to prepare in the study, and in advance, for the dissection of each succeeding day. Practical suggestions as to the best method of demonstrating, precede the descriptions of the various regions and parts of regions. Illustrations have been omitted, for the reason that they add to the expense of a book, often without enhancing its real value, and from the belief that they are liable to great abuse, by distracting attention from the descriptive text to the numbered references, the simple verification of the latter taking the place of the full information only to be obtained from the former.

BOSTON, *Nov.* 1858.

CONTENTS.

General Rules to be observed in Dissecting xi

PART FIRST.

ANATOMY OF THE HEAD AND NECK.

DISSECTION	PAGE
I. External Ear	13
Frontal and Orbital Region	15
II. Facial Region	19
Facial Arteries	22
Facial Nerves	23
Parotid Gland and Region	24
III. Dura Mater and Sinuses	26
Arteries and Muscles of the Orbit	31
IV. Cranial Nerves at their Exit from the Skull . . .	33
V. Superficial Cervical Region	38
VI. External Carotid Artery	44
Submaxillary Region	46
VII. Pterygo-maxillary Region	48
Articulation of the Lower Jaw	48
Deep Cervical Region	51
VIII. Sterno-clavicular Articulation	54
Base of the Neck	55
IX. Pharynx	60
Palatine Region	62
Otic and Meckel's Ganglia	64
Nasal Fossæ	65
X. Tongue	66
Larynx	68
XI. Prevertebral Region	72
Ligaments of the first two Vertebræ	74
XII. Anatomy of the Eye	75

DISSECTION	PAGE
XIII. Membranes and Vessels of the Encephalon	79
Origins of the Cranial Nerves	82
Medulla Oblongata and Pons Varolii	83
Base of the Cerebrum	85
Upper surface and Interior of the Cerebrum	86
Cerebellum	92
XIV. The Internal Auditory Apparatus	94
Middle Ear	95
Internal Ear	97

PART SECOND.

ANATOMY OF THE UPPER EXTREMITY, THORAX, AND BACK.

I. Pectoral and Deltoid Region	100
Axilla	102
II. Front of the Arm	105
Bend of the Elbow	109
III. Sternal Region	110
Ligaments of the Sternum and Costal Cartilages	111
Anterior Mediastinum	112
Heart	115
IV. Posterior Mediastinum	120
Lungs	124
V. The Back and Posterior Cervical Region	125
Spinal Cord and Membranes	134
VI. Scapular Region	137
Back of the Arm	139
VII. Front of the Forearm	140
VIII. Back of the Forearm and Hand	146
IX. Palm of the Hand	150
X. Ligaments of the Ribs, Spine, and Upper Extremity	156

PART THIRD.

ANATOMY OF THE ABDOMEN AND LOWER EXTREMITY.

I. Parietes of the Abdomen	162
Anatomy of Inguinal Hernia	168
II. Visceral Cavity	172
Peritoneum	173
Ducts, Vessels, and Nerves of the Abdominal Cavity	176

DISSECTION	PAGE
III. Intestinal tube	183
Spleen	186
Pancreas	187
Liver	187
IV. Supra-renal Capsules	191
Kidneys	191
Diaphragm	193
Superficial Femoral Region	195
Anatomy of Femoral Hernia	196
Lumbar Plexus	199
V. Anatomy of the Perineum	201
VI. Interior of the Pelvis	206
VII. Rectum	212
Bladder	213
Vesiculæ Seminales and Prostate	214
Penis	215
Testes	217
Female Organs of Generation	219
VIII. Anterior Femoral Region	223
IX. Gluteal Region	229
X. Posterior Femoral Region	233
Popliteal Space	235
XI. Front of the Leg and Dorsum of the Foot	237
XII. Back of the Leg	241
XIII. Sole of the Foot	245
XIV. Ligaments of the Pelvis and Lower Extremity	249

Peculiarities in the Anatomy of the Fœtus 256
Important Anatomical Landmarks and Points, capable of being studied without dissection, or upon the living subject . 260

A FEW GENERAL RULES

TO BE

OBSERVED IN DISSECTING.

The necessary incisions to expose any part, while they follow as nearly as may be the direction of, and penetrate through to, the muscular fibres underlying them, should be arranged in such a manner as to preserve the skin in the largest possible flaps, no other covering of the dissection, when temporarily abandoned, preserving it in an equally good condition. The skin is to be removed only so far as freedom of dissection, in any given region, requires. The fingers should take the place of the forceps as soon as the skin is sufficiently raised, as they stretch it more evenly and over a greater extent of surface.

In dissecting muscles, the fibres must always, if possible, be made tense. The sheath of a muscle should, as a rule, be detached with the skin and never be left behind for subsequent removal. The knife should operate in long sweeps, using its convexity and not its point, following the direction of the fibres, the dissection of any one bundle of which should be completed in its whole length before a second is commenced. The forceps must never seize the muscular fibres, as by so doing they are torn and made to present a ragged appearance. The deep surface of a muscle is to be cleaned as thoroughly as its superficial; the tendinous extremities are to be isolated with special pains that their points of attachment may be precisely studied.. When muscles are to be divided, the section should always, if practicable, be made in their central portion, and never at one of their attachments; the two ends may then at any time be reapplied, and the deep relations, especially to articulations, can thus be better appreciated.

Arteries should be dissected with a pointed knife, and, so far as is practicable, from the trunk toward the branches, and the branches from their origin toward their termination. The forceps should steady the artery by seizing its sheath, and not the vessel itself; if the forceps are what they should be, this rule need never be violated; inferior forceps are a greater hinderance to neat dissecting than dull scalpels. Nerves are to be put upon the stretch with hooks, and stripped of their cellular sheath with care, for they are liable to be unintentionally divided.

The whole subject, when not in use, should be covered, and each part, with the integument belonging to it replaced, should be wrapped around by a bandage *dampened, not wet*, with water. Parts intended to be preserved for the study of the ligaments must be kept merely moist enough to prevent drying; anything like maceration giving all the tissues a uniform opacity which renders the distinction of the ligaments at once difficult and unsatisfactory.

Slowness, without unnecessary delay, and industrious, systematic application, completing whatever is commenced, before beginning elsewhere, are requisites for good dissecting. Neatness and cleanliness, both as to the table and the part being dissected, as well as the hands of the dissector, will contribute, not only to personal comfort, but to the avoidance of diarrhœa and the dangers which sometimes follow dissecting wounds.

If a wound is received while dissecting, it should immediately be held beneath the running water-faucet and thoroughly washed; the bleeding must be encouraged, and any matter introduced drawn out by sucking the part. It is a questionable practice to cauterize such wounds, but if it be done, a saturated *solution* is the only form in which nitrate of silver should be used.

PART FIRST.

ANATOMY OF THE HEAD AND NECK.

The dissection of the regions included under the general term of "the head" is considered the most difficult of any in the body. Comprising more important structures within smaller limits than any other part, and the osseous structure of the skull constituting a large portion of their bulk, it presents mechanical obstacles not easily surmounted. It is hardly possible to obtain an idea of all the different component parts in one dissection, as can be done in other regions of the body; the muscles and arteries may be dissected upon one side of the head, and the other reserved for the nerves; a special part is, however, desirable for the preparation of the nerves; it should not be injected, and it is well to preserve it in spirit, so slowly is their dissection accomplished.

DISSECTION I.

EXTERNAL EAR.

The dissection of the head is usually commenced by an examination of the muscles of the ear; for this purpose the head rests upon its side, and the hair should be shaved from the scalp. A hook inserted into and drawing the margin of the ear successively, backward to dissect the attrahens, downward to dissect the attollens, and forward to dissect the retrahens, brings in turn the tendons of the three auricular muscles into relief; the skin over these, with the cellular tissue beneath, being cautiously removed, the delicate muscular fibres, of which they are composed, may be demonstrated. The dissection is sometimes made puzzling to the beginner by the pale color and small size of the aural muscles.

The muscles which attach the ear to the side of the head are called its *extrinsic* muscles, and are three in number; the special muscles of its cartilage, or pinna, are called *intrinsic*.

The ATTRAHENS AUREM is the most anterior of the extrinsic muscles; it arises, pale and indistinct, from the epicranial aponeurosis, just above the zygoma, and its fibres are directed backward to be inserted into the anterior part of the cartilaginous rim of the ear. It is often wanting, or its place is supplied by the anterior fibres of the attollens aurem.

The ATTOLLENS AUREM is a better-marked muscle than the preceding; it arises, fan-shaped, from the epicranial aponeurosis, on the side of the head above the ear, and is inserted into the upper and anterior part of the concha of the ear.

The RETRAHENS AUREM arises by several separate slips from the mastoid process of the temporal bone, and is inserted into the posterior surface of the concha. This is usually the largest and best marked of the aural muscles.

In dissecting these muscles the *anterior auricular artery*, a branch of the temporal, and the *posterior auricular*, a branch of the external carotid, or sometimes of the occipital, will be seen; they are distributed as their names indicate.

The following nerves should also be sought for in this connection. The *auricularis magnus nerve*, an ascending branch of the anterior cervical plexus, is distributed to the back of the ear; the *posterior auricular*, a branch of the facial nerve, accompanies the artery of the same name, and is distributed to the ear and occipital region. The *occipitalis major*, a posterior branch of the second cervical nerve, emerging from the deep muscles of the back of the neck, and accompanying in part of its course the occipital artery, sends a branch to the ear. The *occipitalis minor nerve*, another ascending branch of the anterior cervical plexus, is placed midway between the last named and the posterior auricular, with both of which it communicates; it supplies the attollens aurem muscle and the integument.

The EXTERNAL EAR is described as consisting of the pinna or auricle, and the meatus. The *pinna*, or projecting part of the ear, consists of a number of folds and hollows, which have been named as follows: the external folded margin is called the *helix;* the elevation which runs parallel to it, the *anti-helix;* the process projecting over the meatus is the *tragus;* opposite to this is a prominence called the *anti-tragus;* the dependent portion of the ear is called the *lobule;* the central depression around the meatus is the *concha;* the space between the helix and the anti-helix, the *fossa innominata;* this terminates anteriorly in a triangular depression called the *scaphoid fossa*. The *meatus* is the cartilaginous canal leading from the pinna to the *tympanum*, a thin, transparent membrane at the bottom of this canal, which separates the external from the middle ear.

The muscles of the pinna require a patient dissection; indeed, they are not always to be found, being only rudimentary in man. The integument covering them is very thin and delicate.

The intrinsic muscles of the ear are five in number.

The *major helicis* consists of vertical fibres, to be found on the anterior border of the helix, just above the tragus.

The *minor helicis* is placed upon that part of the helix which extends into the concha; its fibres are arranged obliquely.

The *tragicus* is a vertical bundle of fibres situated on the outer surface of the tragus.

The *anti-tragicus* arises from the outer part of the anti-tragus, and its fibres converge to be inserted into the pointed extremity of the anti-helix.

The *transversus auriculæ* is found on the posterior aspect of the ear, stretching transversely across the depression between the helix and the concha.

On removing all the integument, the pinna of the ear will be found to consist of a single cartilage, presenting the general outlines of the ear; it does not extend into the lobule, and its continuity is broken by notches and fissures. It is firmly attached to the processus auditorius of the temporal bone by a cartilaginous tube called the *meatus auditorius externus*. In the subcutaneous cellular tissue of the meatus are the *ceruminous glands*, which secrete the wax of the ear.

FRONTAL AND ORBITAL REGION.

To dissect the occipito-frontalis muscle, make an incision from the root of the nose backward to the occiput; this should be met at the vertex by an incision made from the ear at right angles to it; another must be carried along the eyebrow, from the root of the nose outward. The flap is to be lifted from the nasal angle, and with great care; for the fibres are traced with difficulty, and the thin plane which they constitute is very adherent to the integument and liable to be cut through or dissected up from the bone upon which it lies; the aponeurotic tendon which expands over the vertex of the skull can hardly fail of being disfigured by "button-holes."

The OCCIPITO-FRONTALIS MUSCLE consists of two portions, a frontal and an occipital; the two being connected by an aponeurosis, which expands over the vertex of the cranium. It may therefore be described as arising from the external part of the superior curved line of the occipital bone, and from the mastoid portion of the temporal bone to form a rounded and flat belly, which terminating in an aponeurosis, passes forward to join the frontal portion lying upon the frontal bone; this, broader and thinner than the occipital, confounds itself with the orbicularis palpebrarum, pyramidalis nasi and corrugator supercilii muscles, and is inserted into the nasal bones and the superciliary ridge of the frontal bone. Anteriorly the muscular bellies of the two sides blend together; posteriorly they are separated by an interval completed by an aponeurotic expansion.

It might seem better, perhaps, to describe the occipito-frontalis as two muscles instead of one single digastric muscle; each being inserted into the epicranial aponeurosis. This would be suggested by its action, it being a muscle of expression, and is the manner in which it is described by some authors. The aponeurosis connecting the two portions of the occipito-frontalis is firmly connected with the skin, though but loosely attached to the pericranium. It expands over the vertex without any separation into lateral parts, and upon the sides of the head is thin and amounts to little more than cellular tissue.

Upon the posterior belly of the occipito-frontalis muscle will be found the *occipitalis major nerve*, a posterior branch of the second cervical nerve, and the *occipital artery*, a branch of the external carotid. The artery is tortuous, and supplies the muscle, integument, and epicranium; it anastomoses with the temporal artery, and sometimes gives off the posterior auricular branch to the external ear. The nerve accompanies the artery only in a part of its course, and is distributed chiefly to the integument.

The PYRAMIDALIS NASI is a slip of the occipito-frontalis passing downward upon the bridge of the nose. Its outline is usually confused, and it is inserted into the nasal bone and the compressor muscle of the nose. Properly, it should be considered as a pillar of origin of the frontal belly of the occipito-frontalis.

The ORBICULARIS PALPEBRARUM surrounds the eye, and is seen by dissecting the integument from the eyelids, upon which it expands, as it also does upon the circumference of the orbit; it is a thin muscular plane, difficult to dissect, owing in a degree to the want of fixedness in the parts upon which it rests. Its fibres are well marked, and it arises from the inner angle of the frontal bone, the nasal process of the superior maxillary, and the tendon of the tarsal cartilages; encircling the orbit, it is inserted into the same point from which it arose, thus making a sphincter muscle. At the inner angle of the orbit certain ascending fibres of the orbicularis, springing from the tendon of the tarsal cartilages, expand fan-shaped, are inserted into the inner half of the eyebrow, and have received the name of *depressor supercilii*.

The CORRUGATOR SUPERCILII lies beneath the upper half of the orbicularis palpebrarum, which must be dissected up in order to expose it. It is usually confounded with the orbicularis, and not always to be satisfactorily separated from it. It arises from the inner part of the superciliary ridge, and is inserted into the under surface of the orbicularis and frontalis muscles, being about an inch in length. Upon the frontal bone, beneath the muscles last dissected, will be found the divisions of the *supra-orbital nerve*, a branch of the fifth cranial pair; these emerge from the supra-orbital notch, and after supplying the muscles which lie over it, are distributed to the epicranium and integument of the forehead. This nerve is accompanied by the supra-orbital branch of the ophthalmic artery, a branch of the internal carotid; emerging at the supra-orbital foramen, it divides and is distributed like the nerve.

The EYELIDS consist of two cartilages, one for the upper and one for the lower lid, that of the upper being the largest; they are covered externally by muscles and integument, and along their free border the eyelashes are inserted; both are semi-lunar in shape, and attached to the edge of the orbit by a membrane called the *ligamentum palpebræ*. A small fibrous band, called the *tendo oculi*, arising from the anterior margin of the lachrymal canal and dividing into processes, one for each cartilage, fixes them internally; a fibrous band also attaches them to the margin of the orbit externally. They are invested internally by a mucous membrane called the *conjunctiva;* this is continuous with that covering the eyeball, from which it is reflected, and on both surfaces is movable and vascular; over the cornea it is thin and transparent, and in the state of health no vessels are traceable in that part. On the ocular surface of the cartilages may be seen numerous parallel and somewhat tortuous lines, indicating the *Meibomian glands*, which open along the free edges of the eyelids. The *lachrymal canal* has an opening at the inner extremity of each eyelid, indicated by a slight prominence, in the centre of which is a small orifice called the *punctum lacrymale*. The canals of the two lids uniting form a common canal, less than one-eighth of an inch in length, terminating in the *lachrymal sac*, which occupies the concave portion of the lachrymal bone. This sac is the expanded upper part of the *nasal duct*, which conveys the tears from the lachrymal canals to the inferior meatus of the nasal fossa (p. 65). In the internal commissure of the eyelids there is a prominent reddish body formed from the conjunctiva, and called the *caruncula lacrymalis*; just external to this is a fold of the conjunctiva, called *plica semilunaris*, in which may sometimes be found a minute cartilage; this is considered as corresponding to the third lid or membrana nictitans of birds.

The eyelids are supplied by the *palpebral arteries*, branches of the ophthalmic, given off near the inner angle of the orbit; the *nasal*, another branch from the same source, emerges above the tendo oculi, and inosculates with the angular branch of the facial artery; the *frontal*, also from the ophthalmic, appears near the same point, and is distributed to the forehead.

The LACHRYMAL GLAND is situated in the hollow of the external angular process of the frontal bone, and admits of examination at this time by dividing the upper eyelid at its centre and at its external angle. It is a thin flattened body, the size of a small chestnut, resembling in its structure the salivary glands; it sends a prolongation down upon the cartilage of the upper eyelid, but its principal portion lies in contact with the periosteum, to which it is held by a few fibrous bands; its inferior surface rests upon the eyeball and the external rectus muscle. It receives a branch of the ophthalmic artery.

To see Horner's muscle, the eyelids must be divided in the middle by a transverse cut and turned toward the nose; the conjunctiva, with the fat and cellular tissue filling the inner angle of the eye, must be dissected away.

The TENSOR TARSI, or HORNER'S MUSCLE, consists of a quadrilateral plane of delicate fibres, closely applied to and arising from the lachrymal bone; it is about four lines wide and six lines long; anteriorly it splits into two bands, which terminate in very delicate tendons, to be inserted into each tarsal cartilage by the side of its lachrymal duct.

DISSECTION II.

FACIAL REGION.

In order to dissect the muscles of the face, the lips and cheeks should be distended by filling the space between them and the teeth with cotton wool or like material; the lips are then to be sewed together; the nostrils may be distended in the same manner. One side should be dissected at a time, the other being preserved to verify the first: for the very feeble development of the facial muscles makes it difficult to distinguish them; the class of individuals who finish their career in the dissecting-room, is not calculated to display the muscles of expression about the nose and mouth, as it does those of the arms or legs developed by constant use.

The integument should be turned downward from the inner side of the orbit; and as the inferior segment of the orbicularis palpebrarum muscle covers in the origin of several facial muscles, it should be dissected up so as to expose them.

The LEVATOR LABII SUPERIORIS ALÆQUE NASI occupies the depression at the side of the nose; it arises from the nasal process of the superior maxillary bone, beneath the orbicularis palpebrarum, and expanding as it descends, is attached to the ala of the nose and the surface of the upper lip, where it becomes confounded with the orbicularis oris. The fibres attached to the ala of the nose are often but a small part of the whole number, and few and faint in appearance.

The LEVATOR LABII SUPERIORIS PROPRIUS is a quadrilateral muscle, covered in by a considerable amount of adipose tissue, and partially obscured by the preceding muscle and the orbicularis palpebrarum; it is a very distinctly characterized muscle, arising from the lower edge of the orbit and the surface of bone beneath, and is inserted into the integument of the upper lip by fibres which become confounded with the orbicularis oris.

Beneath this muscle will be found the branches of the *infra-orbital nerve:* being the terminal filaments of the superior maxillary branch of the fifth pair of cranial nerves; they emerge at the infra-orbital foramen, and, expanding on the side of the nose and upper lip, freely anastomose with branches of the facial nerve. This nerve is accompanied by the infra-orbital vein, and by the terminal branches of the internal maxillary artery, called the *infra-orbital;* these emerge also at the infra-orbital foramen, and are distributed like the accompanying nerve, anastomosing with branches of the facial artery.

The ZYGOMATICUS MINOR MUSCLE is the internal of two slender muscular slips passing from the malar bone to the angle of the mouth; it arises from the face of the malar bone and passes obliquely to the integument of the lip near its angle, where it is inserted, blending with the insertion of the levator labii superioris proprius. This muscle is very often wanting,—and, when present, seems to be a bundle of fibres of the orbicularis palpebræ detached, to pass downward to the angle of the lips.

The ZYGOMATICUS MAJOR MUSCLE arises outside the preceding, from the surface of the malar bone near its external angle, and is inserted into the integument of the angle of the lips, blending with the orbicularis oris. As its name implies, this muscle is considerably larger than its companion.

The LEVATOR ANGULI ORIS arises from the canine fossa of the superior maxillary bone, and is covered in by the levator labii superioris proprius; it is inserted into the angle of the mouth, where it confounds itself with the orbicularis oris and the other muscles converging at that point.

The DEPRESSOR LABII SUPERIORIS ALÆQUE NASI can only be seen by turning the upper lip inside out and dissecting off the mucous membrane on each side of the frenum; it is a small, pale muscle, not easily detected, the fibres of which are confounded with those of the orbicularis oris; it arises from the fossa in the superior maxillary bone just above the incisor teeth, and is inserted into the upper lip and the cartilages of the ala and septum of the nose.

The COMPRESSOR NASI expands in a radiated manner upon the side of the nose; its fibres are very thin and pale, and its origin is covered in by the levator labii superioris proprius; it arises from the canine fossa of the superior maxillary bone, and ends in an aponeurosis covering the cartilaginous part of the nose, joining the tendon of the other side: its precise limits are difficult to establish.

The CARTILAGES OF THE NOSE are five in number, two on each side, and one in the centre, the latter forming the septum of the nostrils. The superior are called the *lateral fibro-cartilages;* they are triangular in shape, and are attached posteriorly to the nasal bones and the nasal process of the superior maxillary bone; anteriorly they are attached to the anterior border of the septum. Below these are the *alar fibro-cartilages;* these are curved in such a way as to form the rim of the nostril; anteriorly they form the apex of the nose, and their inner portions at this part turn backward along the septum nasi. This cartilage has no osseous attachment, but is connected by fibrous tissue with the lateral cartilages and the integument. To the outer curve of the alar cartilage are attached several small cartilages connected with each other by fibrous tissue; these are called the *sesamoid* cartilages. The *septum nasi* is triangular in form and divides the nose into its two nostrils: it is attached above to the nasal bones and lateral cartilages, posteriorly it unites with the vomer, and below with the palate process of the superior maxillary bones.

The ORBICULARIS ORIS is an elliptical-shaped muscle forming a sphincter round the mouth; it has no osseous attachment; its fibres cross each other at the angles of the mouth; those belonging to the upper lip join the lower portion of the buccinator muscle, and those from the lower lip join with the upper part of the same muscle. Beneath this muscle lies the mucous membrane of the lips, and between this and the muscular fibres are the coronary branch of the facial artery, and numerous small, rounded mucous glands, called *labial glands.*

The DEPRESSOR LABII INFERIORIS, or QUADRATUS MENTI, arises from the oblique line of the inferior maxillary bone, and, blending with its fellow of the opposite side, is inserted into the orbicularis oris at the central part of the portion belonging to the lower lip.

The DEPRESSOR ANGULI ORIS, or TRIANGULARIS MENTI, arises more externally from the same oblique line of the inferior maxillary bone, and is inserted into the orbicularis oris near the angle of the lips; its most external fibres will be found continuous with some of those of the zygomaticus major.

Both of the last-described muscles are made up of fibres from the platysma myoides of the neck; and if they have been carefully dissected, the continuity may be traced with the greatest ease.

Beneath the depressor anguli oris will be seen issuing from the mental foramen, the termination of the *inferior dental* branch of the inferior maxillary trunk of the fifth pair of cranial nerves; its filaments supply the muscles and integument of the lower lip and anastomose with branches of the facial nerve. The *inferior dental* branch of the internal maxillary artery, accompanied by its vein, also emerges at the mental foramen, and, communicating with the facial artery, is distributed to the structures covering the lower jaw.

The LEVATOR LABII INFERIORIS is an extremely small muscle, difficult to isolate; it is to be sought from the inside of the under lip, which should be turned down and the frenum divided. The muscle arises from the surface of bone just below the incisor teeth, and is inserted into the integument on the tip of the chin.

The BUCCINATOR MUSCLE arises from the outer surface of the alveolar borders of the upper and lower jaw, as far forward as the first molar tooth; also from a fibrous raphé called the pterygo-maxillary ligament, which intervenes between it and the superior constrictor of the pharynx; its fibres converge anteriorly and become continuous with those of the orbicularis oris. The intersection of the muscles anteriorly should be dissected with special care. At about its centre, the buccinator is perforated by the excretory duct (Steno's) of the parotid gland; this lies partly imbedded in a quantity of fat which occupies the interval between this muscle and the masseter, and care must therefore be taken not to divide it.

The RISORIUS SANTORINI MUSCLE consists of a few fibres of the platysma myoides, varying in their degree of distinctness, which pass transversely inward over part of the buccinator and masseter muscle, to terminate near the angle of the mouth.

FACIAL ARTERIES.

The facial arteries may be made the subject of a special dissection upon one side of the face; or, if the student has been careful to preserve them in connection with the muscles, they may be studied on the side already dissected. The arteries proper of the face are the facial and transverse facial; the former a branch of the external carotid, the latter of the temporal artery.

The FACIAL ARTERY emerges from the neck just anterior to the masseter muscle, where it rests upon the lower jaw, covered in by the platysma myoides; it passes upward obliquely and tortuously: near the angle of the mouth it gives off an *inferior labial* branch, which passes beneath the depressor anguli oris muscle, and is distributed to the lower lip and chin. The *superior* and *inferior coronary* branches are given off separately, or by a common trunk; they supply the upper and lower lip, lying between the orbicularis oris muscle and the mucous membrane, and inosculate with the corresponding branches of the opposite side. The superior coronary sends a branch to the septum of the nose. The facial continues up beside the nose under the name of the *lateralis nasi*, its termination being called the *angular artery*, and anastomoses with the nasal branch of the ophthalmic. The facial artery is apt to be irregular, and is seldom symmetrical on the two sides of the face.

The *facial vein* accompanies the facial artery on its outer side, and, uniting with the temporal vein, terminates in the internal jugular vein.

The TRANSVERSE FACIAL ARTERY is a branch of the temporal artery; it emerges at the anterior border of the parotid gland, and lies beside the parotid duct; it anastomoses with the facial artery, and supplies the muscles and integument. It occasionally arises from the external carotid.

FACIAL NERVES.

If a special part is not to be obtained for the dissection of the facial nerves, they may be examined on a side of the face reserved for that purpose, or, if sufficiently preserved, in connection with the dissection already made. Some of the nerves are concealed by the parotid gland, but a greater part are external to it; the external branches are to be followed out one by one, and, to see those within the gland, it must be removed piece by piece while tracing them backward.

The FACIAL NERVE is a portion of the seventh cranial nerve, and issues from the skull at the stylo-mastoid foramen; it divides near the ramus of the jaw into two divisions, the temporo-facial and the cervico-facial; the *posterior auricular* branch, which was dissected with the external ear, is given off close to the skull, and turns upward in front of the mastoid process to supply the attrahens and attollens aurem and the posterior belly of the occipito-frontalis muscle.

The *temporo-facial* division gives off a large number of filaments, which expand upon the side of the face as temporal, malar, and infra-orbital branches, anastomosing freely with the supra-orbital and infra-orbital branches of the fifth cranial nerve.

The *cervico-facial* division is smaller than the preceding; its filaments are distributed upon the lower part of the face as buccal, supra-maxillary and infra-maxillary branches; the supra-maxillary branches course inward toward the chin, and beneath the depressor anguli oris anastomose with the inferior dental branch of the fifth nerve. The web-like aspect of this network of communicating branches has given them collectively the name of *pes anserinus*.

The *auriculo-temporal* branch of the inferior maxillary trunk of the fifth pair is also partly seen in this dissection; it emerges from beneath the parotid gland and ascends in company with the temporal artery to the side of the head; it communicates with the facial nerve and supplies the integument in front of the ear, the terminal branches being distributed to the epicranial and temporal aponeuroses.

The infra-maxillary branches are situated below the lower jaw, lying beneath the platysma, and ramify as far as the hyoid bone.

PAROTID GLAND AND REGION.

The PAROTID GLAND, the largest of the salivary glands, is an irregularly shaped body made up of lobes and lobules; it lies in front of the ear, and is partly covered by the platysma muscle; it is limited above by the zygomatic arch, behind by the meatus of the ear and the sterno-mastoid muscle; inferiorly, it descends as low as the posterior belly of the digastricus muscle, and its deep surface penetrates in various directions to a considerable depth; anteriorly, it expands upon the side of the face, and a small accessory part, called *socia parotidis*, is prolonged from it over the masseter muscle. Its excretory duct, called the *duct of Steno*, is given off from the anterior portion; it can be traced but a short distance into the substance of the gland itself; the length of the duct is about two inches and a half, and its size is about equal to that of a crow-quill. It is composed of a fibrous and a mucous coat, and perforates the cheek obliquely opposite the second molar tooth of the upper jaw. A line drawn from the meatus auditorius to a little below the nostril would mark the course of the duct in the cheek, and its orifice would correspond to a spot midway between these two limits. This is a point to be remembered in operations on the face. The internal carotid artery and internal jugular vein lie beneath the gland; the external carotid, accompanied by several large veins, passes through its middle, sending off numerous branches; curving from behind forward the facial nerve divides within its substance. Its own vessels and nerves are derived from the external carotid artery and the facial nerve.

The examination of the parotid will demonstrate the danger, if not the impossibility, of completely extirpating this body during life; even in the dead subject it will be found that it is a very nice dissection to remove it neatly and properly. The scissors will be found a useful instrument in this operation, which should be commenced from the external carotid; the branches of this, many of which are small, should all be saved, so that when the gland is removed the artery shall be left with all its offsets.

The MASSETER MUSCLE arises from both surfaces of the zygoma and from the malar bone, and is inserted into the outer surface of the coronoid process, ramus, and angle of the inferior maxillary bone. In shape it is quadrilateral; externally it is covered by a strong shining aponeurosis obscuring its muscular fibres, and which is not to be removed. The direction of the superficial fibres of this muscle is obliquely backward; of its deeper fibres obliquely forward.

The TEMPORAL APONEUROSIS is a brilliant fibrous membrane covering in, and at its upper part adherent to, the temporal muscle: it is attached superiorly to the curved line that limits the temporal fossa, and below to the zygoma by two layers, between which there is a layer of loose fat and cellular tissue.

The TEMPORAL ARTERY ramifies upon the temporal aponeurosis; it is one of the terminal branches of the external carotid, arising within the parotid gland, where it gives off the transverse facial branch (p. 23). Ascending upon the temporal fascia, just above the zygoma, it sends a small branch, called the *middle temporal*, through the fascia to the temporal muscle, and then divides into two branches, anterior and posterior; the *anterior* is distributed in a very tortuous manner to the forehead, where it anastomoses with the supra-orbital branch of the ophthalmic artery; the *posterior* passes backward to the occiput, where it anastomoses with the occipital artery.

The temporal aponeurosis is to be incised around its edges and removed; the zygoma is to be cut through with a chisel just in front of the ear, and the malar bone is to be sawed through in front of the masseter muscle; this muscle being partially detached from the surface of the ramus of the jaw, and turned downward with the portion of the zygoma and malar bone, and a layer of fat and soft cellular tissue removed, the whole temporal muscle will be exposed.

The TEMPORAL MUSCLE arises from the temporal aponeurosis and from the surface of the depression on the side of the skull known as the temporal fossa. Its fibres converge to form a strong, flat tendon, which is inserted into all the inner surface of the coronoid process of the inferior maxillary bone, from its apex to near the last molar tooth.

DISSECTION III.

DURA MATER AND SINUSES.

The removal of the calvaria may be accomplished by breaking through the skull on a line just above the frontal sinuses and the tops of the ears, with the sharp part of a French hammer, or it may be done more neatly with a saw, the track for which has been marked out by the aid of a string tied round the head. The saw is to be carried through the outer table of the bone only, and the inner one fractured by a chisel and mallet; this saves the membranes of the brain from being wounded, but does not leave the bones in so neat a condition for preservation, when that is desired. The sensation communicated to the hand by the saw, and the color of the bone-dust, tell the operator when he has got through the outer table and reached the vascular diploë between the two tables. The application of very considerable force will be required, by either prying or pulling with a hook or chisel introduced into the line of the incision, to effect the detachment of the calvaria from the membranes adherent to it.

The DURA MATER is the most external of the cerebral membranes; it acts as periosteum to the bones of the skull and as a support to the brain; it is rough externally, where it is torn from the calvaria, and especially so along the line of the sutures; parts of it are occasionally left adhering to the detached calvaria. Upon its surface along the median line some small fibrous bodies, called *Pacchionian glands*, will be noticed; they are wanting in infancy and most numerous in old age; depressions in the bone corresponding to them will sometimes be found.

The dura mater should be cut with scissors along the margin of the sawed skull except at the median line in front and behind. The two flaps thus formed may be turned up on to the top of the brain. It will now be seen that its inner surface is smooth and polished; this is due to a serous coat which is the parietal part of the arachnoid, a serous membrane investing the brain and reflected upon the dura mater, and which will be more particularly spoken of hereafter. The dura mater is therefore a fibro-serous membrane. By separating the hemispheres of the brain it will be seen that a process of the dura mater penetrates like a septum between them: this is called the *falx cerebri*. Narrow in front, it is attached to the crista galli of the ethmoid bone; broader posteriorly, it is continuous with a horizontal expansion of the dura mater lying between the cerebrum and cerebellum, called the *tentorium;* its superior border is attached along the vertex of the skull, and the inferior, which is concave and free, nearly reaches the corpus callosum of the brain. The falx sometimes presents perforations of various sizes, or a rarefaction of its fibres, which give it a lace-like appearance; more rarely a solution of continuity, sufficient to allow the hemispheres of the brain to come in contact with each other, has been noticed.

The *superior longitudinal sinus* occupies the convex border of the falx, and may be laid open with the scissors; it extends from the crista galli to the internal occipital protuberance, is triangular in shape, and perforated by numerous small veins; in its interior, transverse fibrous bands, called *chordæ Willisii*, cross it here and there, and occasionally Pacchionian glands are found within the sinus. It terminates posteriorly at the *torcular Herophili*. The *inferior longitudinal sinus* occupies the concave border of the falx; this terminates in the *straight sinus* of the tentorium, which also continues to the torcular Herophili, the common centre of several sinuses, the situation of which corresponds to the internal occipital protuberance. The straight sinus receives the *venæ Galeni*, which come from the interior of the brain.

The brain must now be removed. The head should be allowed to depend as much as possible, and the operation is to be commenced by the division of the anterior attachment of the falx; this is to be raised and thrown backward, but not detached posteriorly; in reflecting this, several veins, entering the superior longitudinal sinus, will necessarily be divided.

The examination of the arteries at the base of the brain is not always easy in the subject which is used for the study of other parts in the same region. If the brain is too soft to be examined with benefit, it is much better to leave the arteries in connection with the skull. The softened cerebral matter left attached to them may be best removed by a stream of water from a syringe or squeezed sponge. In a fresh brain, or one which will harden well, they should be studied in connection with the base of the brain; in that connection they will be described.

The anterior lobes are to be carefully lifted by the fingers, together with the olfactory bulbs from each side of the crista galli. The carotid arteries, optic and motor oculi nerves are to be divided on each side of the processus olivaris, and the pituitary gland lifted out of the sella turcica; this sometimes cannot be done, and then its pedicle must be divided. The hemispheres being supported in the left hand, and gradually allowed to roll out of the cranial cavity, the tentorium is brought into view, and, just in front of its anterior margin, the fourth or trochlearis nerve; this is to be divided, and the tentorium is to be detached from the petrous portion of the temporal bone by cutting as close to its insertion thereto as possible. The fifth nerve is next brought into view; the sixth, small and slight, near the median line of the clivus Blumenbachii; more externally the seventh pair, composed of two parts, the facial and auditory; below this the three trunks of the eighth pair; the upper being the glosso-pharyngeal, the flat band next below, the pneumogastric; and the one ascending from the spinal canal, the spinal accessory. These being divided, as well as the ninth, which is the remaining one, the vertebral arteries are to be cut off close to the spinal cord. The spinal cord is to be cut across, with the spinal nerves on each side, as far down the canal as possible, and the brain will then roll out of the cranial cavity.

The brain should be laid in a basin with the base uppermost; it should be immersed in alcohol to harden and preserve it, and covered with a piece of doubled cotton cloth. The cloth will keep it wet by imbibition, if there is not alcohol enough to cover it.

It is, however, to be borne in mind that the brain softens very soon after death, and that, unless early removed, it may be unfit for dissection. A portion of each hemisphere may be shaved off, in order to give the spirit a better opportunity to penetrate; but even this may not prevent its decomposition in some of its deep-seated parts. The autopsy-room is the best place for the study of the brain, as it can always there be seen in its fresh and naturally firm condition.

The dura mater at the base of the cranial cavity will be found very adherent, on account of the sutures into which it penetrates, and the foramina through which it is continued to form the sheath of the nerves; it is also prolonged downward into the spinal canal where it forms a loose investment of the spinal cord. The *tentorium*, if replaced, will be found to separate the spaces occupied by the cerebrum and cerebellum, and to be attached along the transverse groove of the occipital bone, the sharp edge of the petrous portion of the temporal bone, and to the posterior clinoid processes of the sphenoid.

In the attachment of the tentorium to the occipital bone will be seen the *lateral sinuses;* at the base of the petrous portion of the temporal bone these sinuses turn downward, and pass through the posterior foramen lacerum; just before entering the foramen the inferior petrosal sinus joins the lateral sinus, and the two united become the internal jugular vein.

The *inferior petrosal sinus* is the continuation backward of the cavernous sinus; it passes along the lower border of the petrous portion of the temporal bone, and terminates as above described. The *superior petrosal sinus* is of small size, and lies along the upper border of the petrous bone in the attached portion of the tentorium; it establishes another communication between the cavernous and lateral sinuses, by entering the latter near the base of the petrous bone.

Projecting below the tentorium is the *falx cerebelli*, which separates the two hemispheres of the cerebellum as the falx cerebri does those of the cerebrum; it ends at the foramen magnum, and is attached along the middle line of the occipital bone. In the attached border of this falx may be found two small sinuses, called the *occipital sinuses*, which terminate in the torcular Herophili.

The *cavernous sinuses* are situated on each side of the sella turcica, and are so called from the existence, in their interior, of trabeculæ, like those of cavernous structures in other parts of the body. In the internal wall of the sinus is the internal carotid artery, covered by minute nervous filaments of the carotid plexus of the sympathetic nerve, and crossed by the sixth or abducens nerve; in the external wall of the sinus are the third (motor oculi), fourth (trochlearis), and ophthalmic branch of the fifth or trifacial nerve.

The cavernous sinuses of the two sides are connected by the *circular sinuses*, which surround the pituitary gland; the *transverse*, or *basilar sinus*, which lies upon the basilar portion of the occipital bone, sometimes unites the cavernous and sometimes the petrosal sinuses; anteriorly, they receive the *ophthalmic veins*, which, after collecting the blood from the eye and structures within the orbit, enter the sinus through the sphenoidal fissure.

The PITUITARY BODY should be sought for in the sella turcica and removed for examination. It is a reddish-gray, solid body, closely fixed in its location by the dura mater; it is composed of two lobes, the anterior of which is the largest; and is connected with the infundibulum of the brain by a peduncle, which should be noticed. In the fœtus the pituitary body is hollow, and communicates with the third ventricle through the infundibulum. It was by virtue of this communication that Vesalius believed the fluid of the ventricles was transmitted to the pituitary body, and from it through the sphenoidal sinuses to the nasal fossæ, thus causing the disease called *pituita*, or catarrh; a theory which charlatans still find for their advantage to maintain.

The dura mater is supplied by a number of arteries called meningeal. The *anterior meningeal*, an offset from the ethmoid branch of the internal carotid, is a small artery which enters the skull by a foramen between the ethmoid and frontal bones, and is distributed to the dura mater in that vicinity. The *middle meningeal* artery is a branch of the internal maxillary; it enters the skull by the spinous foramen of the sphenoid bone, and passing into a deep groove in the inferior angle of the parietal bone, spreads over the side of the cranial cavity; the *meningea parva*, from the same source, enters by the foramen ovale, and is distributed to the middle cranial fossa. The *inferior meningeal* arteries are branches from the ascending pharyngeal and occipital, which enter by the foramen lacerum posterius, and, together with the *posterior meningeal*, from the vertebral artery, supply the middle and posterior cranial fossæ.

The *vertebral arteries* will be seen entering the foramen magnum, where they pierce the dura mater. The arteries of the two sides converge in front of the medulla oblongata, and become united in one trunk, the basilar, which will be described with the base of the brain. The vertebral artery gives off many small branches, and one of some size, the *posterior meningeal*, to the cerebellar fossa and the falx cerebelli.

The *internal carotid artery* perforates the base of the skull, at the apex of the petrous portion of the temporal bone; given off from the common carotid in the neck, it ascends to the carotid foramen in the temporal bone, and pursuing a tortuous course in an osseous canal through that bone, enters the cranial cavity. The cranial part of the vessel describes an S-like curve at the side of the sella turcica where it lies in the cavernous sinus; near the anterior clinoid process it gives off the ophthalmic artery, and then, turning upward, divides into branches to supply the brain. As it winds through the cavernous sinus it is surrounded by nervous plexuses derived from the sympathetic nerve of the neck; the branches of these plexuses are very minute, and in an injected subject are not likely to be found; they are called the *carotid* and *cavernous plexuses*, the former being on the outer side at the entrance of the sinus, and the latter close to the root of the anterior clinoid process, which should be cut away to examine it.

ARTERIES AND MUSCLES OF THE ORBIT.

In order to examine the ophthalmic artery and other contents of the orbit, the frontal bone should be sawed down to the orbit at its inner as well as at its outer angle, and the two incisions continued backward with a chisel till they meet near the optic foramen; the bone being turned down over the eye, but not removed, the orbit is exposed, filled with a soft delicate fat. - The muscles and arteries require patience for their dissection; these may be dissected upon one side, and the nerves upon the other. The eye of any large animal, if removed with all the contents of the orbit, permits the verification, on a larger scale, of many of the points about to be described.

The *ophthalmic artery* enters the orbit at the optic foramen on the outer side of the optic nerve; it gives off numerous small branches not always reached by the injection. Its first branch is the *lachrymal;* this lies along the upper border of the external rectus muscle, in company with the lachrymal nerve; it is distributed to the lachrymal gland and eyelids. The *supra-orbital* branch rests upon the levator palpebræ muscle, and passes forward through the supra-orbital foramen, where it divides upon the forehead, and is distributed to the muscles and integument. The *ethmoidal* branches, two in number, pass through the ethmoidal foramina, and are distributed to the dura mater and nasal fossæ. The *palpebral* are given off by a common trunk which divides into two branches near the inner angle of the eyelids, to which they are distributed. The *frontal* branch emerges at the inner angle of the eye, and is distributed to the forehead. The *nasal* also emerges at the inner angle, and divides into two branches, one of which anastomoses with the angular or terminal branch of the facial, and the other under the name of the *dorsalis nasi* is distributed to the nose. The muscular branches supply the muscles, and give off some small twigs to the eyeball, called *anterior ciliary*. The *short ciliary* branches enter the eyeball around the optic nerve; two of these, one on each side, piercing the sclerotic, farther forward, are called the *long ciliary* arteries. The *arteria centralis retinæ* is a very small branch which perforates the optic nerve, and is distributed to the interior of the eyeball.

The dissection of these arteries will have in a measure effected the dissection of the muscles of the orbit.

The LEVATOR PALPEBRÆ is the most superficial of the orbital muscles; it arises from the roof of the orbit in front of the optic foramen and is inserted by a broad tendon into the tarsal cartilage of the upper eyelid.

This muscle is to be divided, and its two ends are to be reflected.

Four straight muscles surround the optic nerve, and are named from their position SUPERIOR and INFERIOR, EXTERNAL and INTERNAL RECTI MUSCLES; with the exception of the external rectus, they arise posteriorly from the circumference of the optic foramen and sheath of the optic nerve by a common attachment; the *external rectus* arises by two heads, the upper one joining the superior rectus, and the lower the inferior rectus, springing also from the lower border of the sphenoidal fissure; between these origins pass the motor oculi (third), the abducens (sixth), and the nasal branch of the ophthalmic nerve, to enter the orbit. The recti muscles are all inserted, at equi-distant intervals, into the sclerotic coat of the eyeball, about a quarter of an inch from the cornea.

The SUPERIOR OBLIQUE MUSCLE, situated in the upper and inner part of the orbit, is a small rounded muscle arising from the inner side of the optic foramen; it ends anteriorly in a tendon which passes through a loop attached to a depression in the frontal bone at the inner part of the orbit, and is thence reflected backward and outward, between the globe of the eye and the belly of the superior rectus, to be inserted by a broad and flat tendon into the sclerotic, between the superior and external recti muscles. From this pulley-like peculiarity the superior oblique is sometimes called the *trochlearis muscle*. The *loop* is a fibro-cartilaginous ring about an eighth of an inch in width, and, as well as the tendon which plays through it, is lined with a synovial membrane.

In order to examine the next muscle, the optic nerve and the recti muscles must be divided near their origin, and the eye gently turned out of its socket, so as to expose its inferior surface.

The INFERIOR OBLIQUE MUSCLE arises from the superior maxillary bone, between the margin of the orbit and the lachrymal groove; it then passes across the orbit, between the globe of the eye and the inferior and external recti muscles, and is inserted into the external and posterior part of the sclerotic. This is the only muscle of the eye which does not arise from the bottom of the orbit.

In dissecting the eye of a sheep, ox, or calf, the student is often puzzled by an additional muscle, of which he finds no description. It is a suspensor or retractor of the eyeball, and, with the exception of man and the apes, appears to belong to all the mammalia; in most instances it is a quadrifid muscle, but in the ruminantia it coalesces into a single infundibuliform muscle, embracing the optic nerve, and attached anteriorly to the sclerotic, behind the cornea.

DISSECTION IV.

CRANIAL NERVES AT THEIR EXIT FROM THE SKULL.

Paragraphs marked with an asterisk can hardly be verified in an ordinary dissection.

Soemmering counts the cranial nerves at their points of exit at the base of the skull as twelve pairs, enumerating each nerve separately. Willis makes them but nine, including in one nerve all the trunks contained in the same aperture of the skull. The latter division is the one adopted.

The cranial nerves, as they enter their foramina, are invested by a process of the dura mater, which constitutes a sheath for them; the pia mater also is prolonged upon them, but both membranes are soon lost in the surrounding tissues. The arachnoid is reflected backward upon the internal surface of the cranial cavity.

The FIRST, or OLFACTORY NERVE, is soft and pulpy, and is often lost in the manipulation of removing the brain; it lies upon the cribriform plate of the ethmoid bone, and toward its extremity assumes a bulbous shape; it sends a large number of fine filaments, through the foramina of the bone beneath, to supply the mucous membrane of the nasal fossæ.

The SECOND, or OPTIC NERVE, diverging from its commissure, as the conjunction with its fellow is called, enters the orbit through the optic foramen. This nerve is large and of a white color; it is accompanied by the ophthalmic artery, and invested with a firm sheath from the dura mater; it has no branches, and continues forward to the eyeball, where it expands into the retina.

The orbital plate of the frontal bone, on the opposite side to that used for the dissection of the muscles and arteries of the orbit, should be broken through with a chisel, in such a way as to make a triangular opening into the cavity, the base of which, an inch and a half wide, should be as far forward as possible; the apex should lay bare the optic foramen, but without injuring the nerves.

As the third and fourth nerves, and a branch of the fifth, lie in the walls of the cavernous sinus, the dura mater which constitutes it must be dissected away. The fifth nerve should also be cleared from the dura mater, and the Gasserian ganglion which lies upon the apex of the petrous portion of the temporal bone in the middle fossa, and into which the fifth nerve soon expands, should be dissected cleanly, and its three large trunks, the ophthalmic, and the superior and inferior maxillary, traced to their points of exit, the sphenoidal fissure, the foramina ovale and rotundum.

The THIRD, or MOTOR OCULI NERVE, pierces the dura mater just in front of the posterior clinoid processes; pursuing its course in the external wall of the cavernous sinus, it enters the orbit through the sphenoidal fissure, passing between the two heads of the external rectus muscle, and divides into two branches. The *superior* branch supplies the superior rectus muscle and the levator palpebræ. The *inferior* branch supplies the internal and inferior rectus and the inferior oblique muscles. It will thus be seen that this nerve supplies all the muscles of the eye, except the external rectus and superior oblique, each of which has a special nerve. These branches are very small, and, lying in the midst of fat, require great patience and considerable skill to dissect.

* The inferior branch of the motor oculi nerve sends a branch of communication to the *lenticular ganglion*, a small rounded body, the size of a pin's head, placed at the back part of the orbit, between the optic nerve and external rectus muscle, and commonly on the outer side of the ophthalmic artery. This ganglion gives off the ciliary nerves, which pierce the sclerotic in company with the short ciliary arteries around the optic nerve.

The FOURTH, or TROCHLEARIS NERVE, is very small, and sometimes puzzling to find. It pierces the dura mater close to the third nerve, and passes along the outer wall of the cavernous sinus to the sphenoidal fissure; entering the orbit, it crosses the levator palpebræ at its origin, and supplies the trochlearis or superior oblique muscle.

The FIFTH, or TRIFACIAL NERVE, the largest of the nine pairs of nerves, consists of two parts or roots. These two portions pass through the tentorium close to the apex of the petrous portion of the temporal bone; the larger division, immediately on reaching the middle fossa of the skull, expands into a flattened ganglion, the *Gasserian;* the smaller division lies beneath the ganglion, and is only seen by turning it over. These two roots are distinct from each other, and the smaller one may be traced onward to the inferior maxillary nerve which it joins outside of the cranium.

The Gasserian ganglion gives off three branches—the *ophthalmic*, and the *superior*, and *inferior maxillary* nerves.

The *ophthalmic nerve*, arising from the upper portion of the ganglion, passes through the outer wall of the cavernous sinus, and enters the orbit through the sphenoidal fissure. It divides into three branches —nasal, frontal, and lachrymal.

* The *nasal* branch passes between the two heads of the external rectus muscle, crosses the optic nerve, and enters the anterior ethmoidal foramen; it then reappears in the cranium at the side of the crista galli, and enters the nasal cavity, in front of the cribriform plate, to be distributed to the mucous membrane and integument of the nose. As the nasal nerve enters the orbit, it sends a branch to the lenticular ganglion. The *frontal* branch passes forward upon the levator palpebræ muscle to the supra-orbital foramen, where it emerges to supply the integument of the forehead. The *lachrymal* is the smallest of the three branches. It passes along the upper border of the external rectus muscle to the lachrymal gland and upper eyelid, to which it is distributed.

The *superior maxillary nerve*, the second or middle division of the fifth pair, passes out at the foramen rotundum, crosses the sphenomaxillary fissure, and enters the orbit by the canal in its floor, in company with the infra-orbital artery, one of the terminal branches of the internal maxillary; they both emerge at the infra-orbital foramen (p. 19), beneath the levator labii superioris muscle, and, forming a plexus with branches of the facial nerve, supply the lower eyelid, upper lip, nose, and cheek.

* This nerve gives off an *orbitar* branch, which, entering the orbit through the spheno-maxillary fissure, divides into two branches; these pass through foramina in the malar bone, and are distributed to the temporal fossa, forehead, and cheek. *Dental* branches also penetrate through small foramina in the tuberosity of the superior maxillary bone to the molar teeth, and from the infra-orbital canal through the lining membrane of the antrum to the anterior teeth.

These branches, as well as that portion of the main trunk lying in the infra-orbital canal, cannot of course be seen without the removal of the eyeball.

The *inferior maxillary nerve* is the longest, and the most inferior in point of position of the three trunks into which the fifth pair divides; as it passes out through the foramen ovale it is joined by the second primary root of the fifth nerve, which lies behind the ganglion; it then divides into muscular, the gustatory, the inferior dental, and the auriculo-temporal branches.

The SIXTH, or ABDUCENS NERVE, pierces the dura mater on the clivus Blumenbachii of the sphenoid bone; it crosses the cavernous sinus, enters the orbit through the sphenoidal fissure, and passes between the two heads of the external rectus or abducens muscle, to which it is distributed.

The SEVENTH PAIR consists of two nerves, the FACIAL and the AUDITORY; the facial is called the *portio dura*, the auditory the *portio mollis*, the former being of a dense, the latter of a soft and pulpy structure. Both of these nerves enter the temporal bone at the meatus auditorius internus.

In order to study these nerves properly, a temporal bone should be immersed in strong alcohol, and afterward softened in hydrochloric acid; it can then be cut with a knife, and its canals followed out.

The *facial nerve* is the smallest of the two trunks; after entering the meatus auditorius internus, it passes through the aqueduct of Fallopius and emerges at the stylo-mastoid foramen; it there divides into two branches, the temporo-facial and the cervico-facial; these have been already described (p. 23). While in the aqueduct of Fallopius, this nerve forms a ganglionic enlargement, called the *intumescentia gangliformis*. A small branch of the Vidian nerve, which passes backward beneath the Gasserian ganglion, enters the hiatus Fallopii to join the intumescentia; this is called the superficial petrosal nerve.

* The *auditory nerve* enters the meatus auditorius internus, and dividing into two branches, the *cochlear* and *vestibular*, is distributed to the internal auditory apparatus.

The EIGHTH PAIR is composed of three nerves, the GLOSSO-PHARYNGEAL, PNEUMOGASTRIC, and SPINAL ACCESSORY; these all pass out at the foramen lacerum posterius.

The *glosso-pharyngeal nerve* passes through a distinct canal of the dura mater in the above-named foramen, and lies at its innermost extremity. That portion of the nerve lying in the jugular fossa presents two gangliform swellings, the superior being called the *ganglion jugulare* and the inferior the *ganglion petrosum*, or *ganglion of Andersch*. The ganglion of Andersch gives off the *tympanic*, or *Jacobson's nerve;* this enters a small bony canal in the jugular fossa, and is distributed to the tympanum, forming the tympanic plexus, which communicates with the sympathetic and with the fifth pair of nerves. One or two other minute branches are also given off from the ganglion or its vicinity. The glosso-pharyngeal nerve is distributed to the base of the tongue, the fauces, and the pharynx.

* The *pneumogastric*, or *par vagum nerve*, passes out of the foramen lacerum posterius in a sheath common to it and the spinal accessory nerve; it is the longest of the three nerves. In the foramen, it has a large ganglion called the *ganglion of the root*, in contradistinction to the ganglion of the trunk, which is formed after it has escaped from the skull.

* The *spinal accessory nerve* passes out of the foramen lacerum posterius with the pneumogastric; it has no ganglion. In the jugular fossa it divides into two branches, one of which sends a few filaments to the upper ganglion of the pneumogastric, with the trunk of which it becomes continuous below the second ganglion, and the other descends to the sterno-mastoid muscle, which it perforates, and to which, as well as the trapezius muscle, it is distributed.

The NINTH, or HYPOGLOSSAL NERVE, consists of two bundles, which pass out at the anterior condyloid foramen by separate orifices of the dura mater. These unite after emerging from the skull, and the nerve is distributed to the muscles of the tongue.

According to the arrangement which this book proposes, the subject is now to be turned over for the dissection of the back: if the student has kept up with his companions, and accomplished the previous dissections, the turning may be done without inconvenience. The muscles of the back of the neck are given with those of the back, in Part Second, Dissection V.

DISSECTION V.

SUPERFICIAL CERVICAL REGION.

The dissection is now transferred to the neck. The head should hang backward by its own weight; the chain hook should be caught into the septum of the nose, and by it the head should be rotated and held to one side, as far as it is possible to do so; this exposes the cervical region, and puts the muscles on the stretch. An incision is made along the median line, from the chin to the sternum, and another, if not already made, along the ramus of the jaw; a third is carried along the clavicle, from the termination of the first, outward. The skin is to be lifted from the angle at the sternum, but as the fibres of the platysma are often very thin and indistinct, and so pale in color as hardly to be distinguished from the fascia between it and the skin, this fascia must be raised with such care as to insure the demonstration of the muscle, however feebly it may be developed.

The PLATYSMA MYOIDES arises from the integument in front of the thorax, below the clavicle; its fibres ascend obliquely forward, uniting upon the median line, when well developed, with those of the other side, and are inserted into the chin, the angle of the mouth, and the integument of the face, being intimately connected with, and in fact helping to form several of the facial muscles.

The platysma is to be removed without disturbing the fascia, or the numerous nerves which lie upon and between it and the sterno-mastoid muscle.

The removal of the platysma-myoides brings the cervical region more fairly into view. It will be seen that it is quadrilateral, and that its boundaries may be indicated in a general way, as, superiorly, the ramus of the jaw and the mastoid process; inferiorly, the clavicle; posteriorly, the edge of the trapezius muscle; and anteriorly, the median line of the neck.

The *cervical fascia* varies in distinctness in different subjects; it surrounds the neck, and is stronger in front of than behind the sterno-mastoid muscle, which it encases. The external jugular vein lies upon it, and perforates it at its lower part, and the branches of the anterior cervical plexus of nerves lie partly upon and partly beneath the fascia.

The EXTERNAL JUGULAR VEIN is formed by the union of the posterior auricular and temporo-maxillary veins, veins of the integument and of the zygomatic and pterygoid fossæ. The external jugular vein is of variable size; it descends the neck, following a line from the angle of the jaw to the middle of the clavicle, crossing the sterno-mastoid muscle, and penetrates the fascia, just at the side of the outer border of the clavicular portion of that muscle, terminating in the subclavian vein. In its course it is joined by the veins which accompany the supra-scapular and posterior scapular arteries in the posterior part of the neck. The *anterior jugular vein*, formed by a series of small branches, collects the blood from the front of the neck, and descending along the anterior border of the sterno-mastoid muscle, sometimes enters the external jugular, and sometimes passes beneath the sterno-mastoid, to join the internal jugular or subclavian vein.

The cervical fascia is to be removed in dissecting the nervous branches which ramify in this region; for the most part these emerge behind the sterno-mastoid, and are to be followed to their origin beneath that muscle so far as may be, without dividing it. The inframaxillary branches of the cervico-facial division of the facial nerve (p. 23) will be found beneath the platysma, between the inferior maxilla and the hyoid bone.

The anterior branches of the four upper cervical nerves communicate with each other by loops, and these loops, together with their branches, constitute the *cervical plexus;* emerging from under the posterior border of the sterno-mastoid muscle, and covered in by the platysma, it is distributed to the muscles and integument. Its branches are divided into ascending, descending, and deep.

The *ascending branches* are three in number:—

    Superficialis colli,
    Auricularis magnus,
    Occipitalis minor.

The *superficialis colli nerve*, coming from the second and third cervical nerves, emerges behind the posterior border of the sterno-mastoid, at about its middle, crosses it in a direction obliquely upward, and divides into ascending and descending branches, which are distributed to the front of the neck.

The *auricularis magnus nerve* comes also from the second and third cervical nerves, and emerges at the posterior border of the sterno-mastoid, on the superficial surface of which it ascends, in close relation with the external jugular vein, to the parotid gland, where it divides into two branches, the anterior being distributed to the external ear, parotid gland, and cheek; the posterior, crossing the mastoid process, supplying the back part of the external ear and the integument in its neighborhood.

The *occipitalis minor nerve* arises from the second cervical nerve, and will be found lying upon the upper part of the posterior border of the sterno-mastoid muscle. It is distributed to the integument of the occipital region, and anastomoses with the occipitalis major, auricularis magnus, and posterior branches of the facial nerve.

The *descending branches* of the cervical plexus come from the third and fourth cervical nerves, and, descending between the sterno-mastoid and trapezius muscles, are named *acromial* and *clavicular*, being distributed to the integument and muscles of the shoulder and anterior and upper part of the thorax. The *deep* branches will be described hereafter. They are chiefly muscular, or connecting filaments with the pneumogastric, sympathetic, or hypoglossal nerves. The *communicans noni* and the *phrenic* are the most important.

The STERNO-MASTOID MUSCLE is the large and prominent muscle which characterizes the cervical region; it is encased by the cervical fascia, and is crossed superficially by the external jugular vein and the branches of the cervical plexus. It arises by two heads, separated by an elongated interval; one, narrow, from the upper bone of the sternum; the other, broader, from the sternal third of the clavicle; the extent of the clavicular attachment varies, and in some bodies may reach even to the trapezius muscle. These two heads unite at about the middle of the neck in a rounded belly, which is inserted into the mastoid process, and by a broad and thin aponeurosis into the superior curved line of the occipital bone.

The OMO-HYOID MUSCLE is a small muscle which traverses the neck diagonally, in a direction crossing that of the sterno-mastoid, beneath which it lies; it is composed of two bellies, united in the middle by a tendon of variable length; only one portion of the muscle can be well seen in the present stage of the dissection. It arises from the scapula at the outside of, and sometimes in part from, the transverse ligament stretching across the supra-scapular notch; it then passes forward, obscured from sight by the clavicle; behind the sterno-mastoid, the scapular portion terminates in a tendon which plays through a loop formed by the deep cervical fascia, this loop being attached to the cartilage of the first rib. From this intervening tendon commences another belly, which pursues a direction upward and forward, to be inserted into the hyoid bone at the point of union between its body and the greater cornu. Occasionally, one of the bellies of this muscle is wanting, and the whole muscle may be absent.

The sterno-mastoid and the omo-hyoid muscles divide the quadrilateral cervical region into "triangles," convenient for purposes of surgical description; thus, the sterno-mastoid passing obliquely across the neck, that portion bounded by the median line in front, the jaw above and the sterno-mastoid behind, is called the *great anterior triangle;* that bounded by the sterno-mastoid in front, the clavicle below and the trapezius posteriorly, the *great posterior triangle.* These triangles are each subdivided by the omo-hyoid muscle; the hyoid portion divides the anterior triangle, the space below it being called the *inferior carotid triangle;* that above it being again divided by the digastricus muscle into the *superior carotid triangle* and the *submaxillary triangle.* The scapular portion divides the posterior triangle into two smaller spaces, that above the belly of the muscle being called the *occipital triangle* and that below it the *subclavian triangle.*

The sterno-mastoid muscle may now be divided and its two ends reflected.

In reflecting the lower half of the sterno-mastoid muscle, its outer border will be seen to correspond with the outer border of the scalenus anticus muscle. As the position of the last-named muscle at its insertion to the first rib is of importance in connection with the operation of ligature of the subclavian artery, this relation is a valuable landmark to recognize.

In reflecting the upper half of the sterno-mastoid muscle, its posterior surface should be examined for the *spinal accessory nerve;* this nerve is one of the eighth pair of cranial nerves (p. 37); after emerging from the foramen lacerum posterius, it becomes connected by a branch of considerable size with the pneumogastric nerve, and then continues onward to perforate the sterno-mastoid muscle at its under surface, after passing through which, and being joined by branches of the cervical plexus in the occipital triangle, it is distributed to the trapezius muscle.

It is presumed that the carotid artery has not yet been exposed, but that it still remains covered with its sheath. Its *sheath* is a portion of the deep cervical fascia which invests it and also the internal jugular vein. Upon this sheath may be seen a small nerve called the *descendens noni;* it is a branch from the hypoglossal, one of the cranial nerves, which crosses the neck in a transverse direction just above the hyoid bone. The descendens noni forms a loop with a deep branch from the second or third nerve of the cervical plexus, called the *communicans noni*. The descendens noni is sometimes found within the sheath instead of upon it.

Upon opening the carotid sheath, the relation of the parts is to be carefully observed. The artery and vein are separated from each other by a thin septum derived from the sheath. The vein is the internal jugular vein, and lies upon the outer side of the artery. Behind and between the artery and vein will be found the pneumogastric or par vagum nerve. Upon the inside of the artery the trachea will be seen, and between the trachea and artery a medium-sized nerve destined to the larynx, and called the recurrent laryngeal. Between the artery and the transverse processes of the cervical vertebræ, on which it rests, may be found that part of the sympathetic nerve which connects the cervical ganglia. Some small branches of the sympathetic nerve, being cardiac branches, should also be noticed in connection with this view of the parts.

The *common carotid artery* is of large size and uniform calibre, and from its origin until its division opposite the upper border of the thyroid cartilage, gives off no branch unless it be a small muscular twig. Its course is indicated by a line drawn from the centre of the interval between the mastoid process and the angle of the jaw to the sterno-clavicular articulation; this line corresponds to the anterior border of the sterno-mastoid muscle, which is called the guide to, or the *satellite* of, the artery. At the apex of the triangle formed by the anterior belly of the omo-hyoid and the anterior border of the sterno-mastoid muscles is the "point of election" for placing a ligature upon the common carotid artery.

Having established these various relations, the dissection may be continued by the examination of the muscles lying upon the trachea.

The STERNO-HYOID MUSCLE lies at the side of the median line of the neck, being covered in at its lower part by the sterno-mastoid, and is a thin ribbon-like muscle about an inch in width, separated from its fellow of the other side by a slight cellular interval. It arises from the internal surface of the first bone of the sternum by a flat muscular origin, and is inserted into the lower border of the body of the hyoid bone. It is occasionally marked by transverse tendinous intersections. It lies upon the sterno-thyroid and thyro-hyoid muscles.

The sterno-hyoid is to be divided and its two ends are to be reflected. In separating the sterno-hyoid and the thyro-hyoid muscles on the median line the student should observe their relations with the trachea and the isthmus of the thyroid gland, these being parts concerned in the operation of tracheotomy.

The STERNO-THYROID MUSCLE arises also from the thoracic surface of the first bone of the sternum, but lower down than the preceding muscle: it is inserted into the oblique line of the thyroid cartilage of the larynx; it is broader than the sterno-hyoid, and, like that, is occasionally marked by tendinous intersections.

The THYRO-HYOID MUSCLE is a short muscle arising from the oblique line of the thyroid cartilage and inserted into the lower border of the body and cornu of the hyoid bone. The separation between this muscle and the sterno-thyroid is not always distinct; normally they are separated by a tendinous interval at the point where they are attached to the oblique line of the thyroid cartilage, but it not unfrequently happens that even this is not sufficiently marked to be apparent.

The sterno-thyroid muscle is to be divided, and its ends reflected, so as to expose the thyroid body.

Between the common carotid arteries and upon the trachea lies the THYROID BODY; this is a dark red and vascular organ composed of two lobes, one on each side of the trachea and larynx, connected by a narrow portion called the *isthmus*, and which lies across the upper two or three rings of the trachea; the lobes are triangular in shape, their bases being directed downward. It is well supplied with arteries, receiving one from each external carotid, called the superior thyroid, distributed to the upper part of the lobes, and one from each thyroid axis of the subclavian, called the inferior thyroid, which supplies the lower part of the lobes; occasionally there is a middle thyroid artery sent to it from the arteria innominata; all these arteries anastomose freely with each other. When the middle thyroid is present, it usually lies upon the trachea in the median line, and may be a source of embarrassment in the operation of tracheotomy. A small muscle called the *levator glandulæ thyroideæ* is sometimes found connected with the upper border of one lobe or with the isthmus, and attached to the hyoid bone; it is said to be most frequent on the left side, and its place is sometimes supplied by a small lobule of glandular tissue, which is then called the *pyramid*, or *middle lobe*.

DISSECTION VI.

EXTERNAL CAROTID ARTERY.

The dissection of the external carotid artery is to be undertaken by following out from the main trunk those branches which have been partly dissected in preparing the parts already examined, as well as those which have not yet been alluded to, but are now to be described.

Opposite the upper border of the thyroid cartilage or a little higher, the common carotid artery divides into two large trunks, the external and internal carotid branches. At first, the external carotid lies upon the inner side, nearer the middle line of the body than the internal carotid, their distinctive names having reference, not to their relative position, but to their destination to parts nearer or more remote from the surface; it soon, however, becomes superficial to the internal carotid and divides into numerous branches. The internal carotid may be distinguished by a peculiar fusiform dilatation at its commencement; sometimes this dilatation is very marked, forming an abrupt rounded distension of the vessel.

The *external carotid* is crossed by the stylo-hyoid and digastricus muscles, and by the hypoglossal nerve, and is imbedded for a part of its course in the parotid gland; between the angle of the jaw and the mastoid process it terminates by dividing into the internal maxillary and temporal arteries. Its branches are the

Superior thyroid,	Posterior auricular,
Lingual,	Ascending pharyngeal,
Facial,	Temporal,
Occipital,	Internal maxillary.

The *superior thyroid artery* descends, passing beneath the omo-hyoid, sterno-thyroid, and sterno-hyoid muscles to the thyroid body, to the superficial surface of which it is distributed, and where it anastomoses with its fellow of the opposite side. It sends offsets to the hyoid region and larynx, under the name of *superior hyoid* and *inferior laryngeal* branches.

The *lingual artery* passes obliquely forward beneath the hyo-glossus muscle. In that part of its course which is parallel to the os hyoides, the hyo-glossus muscle separates it from the hypoglossal nerve, the latter being the more superficial. The lingual artery supplies the tongue, and is continued forward to the tip of that organ under the name of the *ranine* artery.

The *facial artery* arises above the lingual, sometimes from a common trunk with it, and is directed upward over the lower jaw to the face. It passes beneath the digastric and stylo-hyoid muscles and becomes imbedded in the submaxillary gland. In its cervical portion it gives branches to the pharynx, tonsils, and submaxillary gland. The *submental* branch arises from the portion within the gland, and passes forward upon the mylo-hyoideus muscle to the anterior belly of the digastricus, where it terminates in branches, some of which turn upward and reach nearly to the lower lip. Before crossing the jaw, which it does close to the anterior inferior angle of the masseter muscle, the facial artery is tortuous, and continues so throughout the rest of its course (p. 22).

The *occipital artery* is a large branch destined to the posterior part of the head; it passes outward beneath the posterior belly of the digastricus, part of the parotid gland, the sterno-mastoid and trapezius muscles; it crosses the jugular vein and the spinal accessory nerve, and the hypoglossal nerve curves around it near its origin. Near the middle line of the occipital bone the artery turns upward, passing through the fibres of the upper part of the trapezius muscle, becomes superficial, and is distributed to the occiput, anastomosing with its fellow, with the posterior auricular and with the temporal arteries. It gives off a small branch, the *inferior meningeal*, which ascends with the jugular vein through the foramen lacerum posterius to the posterior fossa of the base of the skull. A large but irregular branch, the *princeps cervicis*, descends the neck between the complexus and semi-spinalis colli muscles and inosculates with the profunda cervicis, a branch of the subclavian artery.

The *posterior auricular artery* ascends between the ear and the occipital bone, and is distributed by two branches to the external ear and side of the head. It sends a small twig, called the *stylo-mastoid*, through the stylo-mastoid foramen to the internal ear. This artery is sometimes an offset from the occipital.

The *ascending pharyngeal artery* arises at the point of bifurcation of the common carotid, and is very apt to be destroyed in the dissection. It is of small size, and ascends between the internal carotid and the pharynx. It divides into two branches, the *inferior meningeal*, which enters the cranium through the foramen lacerum posterius, and is distributed to the membranes of the posterior fossa of the skull, and the *pharyngeal*, which is distributed to the mucous membrane of the pharynx and the soft palate.

The *temporal artery* is the terminal continuation of the external carotid; it passes up between the ear and the articulation of the jaw, through the substance of the parotid, and upon the temporal fascia divides into two branches, *anterior* and *posterior temporal*, which ramify on the front and the side of the head. This artery gives off *parotidean* branches to the parotid gland; the *anterior auricular* to the external ear; the *transverse facial* to the muscles of the face, crossing the cheek transversely beside Steno's duct, and anastomosing with branches of the facial (p. 23); the *orbitar* to anastomose with the palpebral arteries, and the *middle temporal*, which perforates the temporal fascia just above the zygoma, and supplies the temporal muscle.

The internal maxillary will be described hereafter.

SUBMAXILLARY REGION.

The dissection of the carotid artery will have exposed and prepared a number of parts situated below the inferior maxillary bone and between it and the hyoid bone.

The DIGASTRICUS MUSCLE, lying above the hyoid bone and connected with it, is composed of two rounded bellies connected by a central tendon, the central tendon being the part attached to the hyoid bone; these two bellies form an obtuse angle with each other; the posterior one arises from the digastric fossa of the temporal bone and is consequently covered in by the mastoid portion of the sterno-mastoid muscle and by the parotid gland. The anterior belly, closely connected, though not united with the corresponding part of the same muscle on the other side of the neck, arises from the side of the symphysis of the lower jaw; these two portions are attached by their central tendon to the body and greater cornu of the os hyoides. The tendon of the muscle is held in its place by a strong fascia and by fibres of the stylo-hyoid muscle which surround it. The posterior belly of this muscle crosses the carotid vessels, and along its lower border will be found the occipital artery and the hypoglossal nerve. The facial nerve sends a branch to this muscle soon after its exit from the stylo-mastoid foramen.

The STYLO-HYOID MUSCLE is in close connection with the preceding; it arises from the styloid process of the temporal bone, and passes down behind and to the inner side of the posterior belly of the digastricus; at its lower part it splits, and allows the digastric tendon to pass through its substance; it is inserted into the os hyoides at the union of its body and cornu. It is sometimes wanting. The facial nerve sends a branch to this muscle also, soon after it emerges from its foramen.

The anterior belly of the digastricus is to be removed, and the submaxillary gland freed from extraneous cellular tissue, and loosened from its attachments.

The SUBMAXILLARY GLAND is a salivary gland, next in size to the parotid, which it resembles in general structure; it lies above the digastricus and upon the mylo-hyoid muscle; it is partly concealed by the lower jaw, though it descends a variable distance down the neck; it is traversed by the facial artery, which distributes numerous small branches throughout its substance; its duct, called *Wharton's duct*, may be seen issuing from its posterior part, and curving around the posterior border of the mylo-hyoid muscle, passes between the hyo-glossus and genio-hyo-glossus muscles and beneath the sublingual gland, to open at the side of the frenum of the tongue.

The MYLO-HYOID MUSCLE arises from the mylo-hyoid ridge of the inferior maxillary bone, and its posterior fibres pass forward to be inserted into the body of the os hyoides; the anterior fibres join with those of the muscle of the other side, forming a sort of raphé along the median line; it is triangular in shape, and with its fellow makes the floor of the mouth, the two muscles stretching across the interval between the two sides of the lower jaw. The *mylo-hyoid* branch of the inferior dental nerve ramifies upon its cutaneous surface, as well as a twig from the internal maxillary artery, which accompanies that nerve. The *submental* branch of the facial artery also ramifies upon this muscle.

The mylo-hyoid muscle is to be carefully removed or turned down toward the hyoid bone.

The GENIO-HYOID MUSCLE lies close to the median line, and arises from the inside of the symphysis of the lower jaw; it is inserted into the centre of the body of the hyoid bone, in close apposition with the muscle of the other side, with which it is often united.

The hyo-glossus and genio-hyo-glossus will be found described in connection with the tongue.

If the submaxillary gland has been preserved, Wharton's duct may be seen resting upon the hyo-glossus muscle. The *hypoglossal nerve* crosses that muscle, and in this part of its course gives off the *descendens noni* branch to the sheath of the carotid vessels (p. 41). By drawing the os hyoides downward the *gustatory branch* of the inferior maxillary nerve will also be found resting upon the hyo-glossus muscle. The *submaxillary ganglion*, a small reddish body, is in close connection with the gustatory nerve, and lies just above the upper border of the submaxillary gland; it is extremely difficult to find. The *chorda tympani*, a branch of the facial nerve which joins the gustatory nerve near the submaxillary ganglion, may also be seen at this point of the dissection.

DISSECTION VII.

PTERYGO-MAXILLARY REGION.

The parts beneath the ramus of the jaw are of difficult dissection; and the numerous important structures crowded into this space, are only by patience eliminated from their apparent confusion, or preserved in sufficient integrity to permit their examination. The zygoma having been already divided at its two ends, and, with the masseter, turned downward, the next step is to saw through the inferior maxillary bone below its neck, and again from the last molar tooth to just below its angle; this fragment, with the temporal muscle attached to the coronoid process, is then to be turned outward, and both in reflecting it, and in sawing it through, the greatest caution is to be observed to divide nothing accidentally upon the inside of the bone, the internal maxillary artery and the inferior dental artery and nerve being in close apposition to it. After this piece of bone has been everted, and a little cellular tissue cleared away, the pterygoid muscles will appear, the external directed toward the condyle of the jaw, and the internal toward its angle. The coronoid process should be examined, to see the extent to which the temporal muscle is attached to its inner surface (p. 26).

The EXTERNAL PTERYGOID MUSCLE arises by two heads from the greater wing of the sphenoid bone, below the crest, and from the outer surface of the external pterygoid plate; its fibres converge to be inserted in front of the neck of the inferior maxilla; its separation into two heads is not always apparent; when present, the second head is inserted into the inter-articular fibro-cartilage of the temporo-maxillary articulation. The internal maxillary artery rests upon this muscle.

The INTERNAL PTERYGOID MUSCLE arises from the pterygoid fossa and from the inner surface of the external pterygoid plate, and is inserted into the angle and inner surface of the ramus of the lower jaw; its fibres follow the same direction as those of the masseter muscle externally, and from this fact, as well as from the correspondence of their insertions, it has sometimes been called the *internal masseter muscle*.

ARTICULATION OF THE LOWER JAW.

In the TEMPORO-MAXILLARY ARTICULATION the condyle of the lower jaw is received into the glenoid fossa of the temporal bone, and is held in place by three ligaments.

The *external lateral ligament* is a short, stout band of

fibres, passing obliquely backward from the tubercle of the zygoma to the outer side of the neck of the lower jaw.

The *internal lateral ligament* is a membranous expansion from the spinous process of the sphenoid bone to the margin of the dental foramen; the internal maxillary artery and dental nerve pass between this ligament and the jaw.

The *stylo-maxillary ligament* extends from the styloid process of the temporal bone to the inside of the angle of the jaw.

These two latter hardly merit the name of ligaments, as they in no way serve to consolidate the articulation; the first being but a membranous protection to the vessels, and the second an aponeurotic surface of origin of the styloglossus muscle.

An *inter-articular fibro-cartilage* is seen on opening the joint; it is elliptical and biconcave in shape, and sometimes perforated in the centre: externally it is attached to the external lateral ligament, and internally to the external pterygoid muscle.

Two *synovial membranes* are present, forming a shut sac above and below the inter-articular cartilage.

The condyle of the jaw, with the external pterygoid muscle attached, is now to be dislocated and drawn forward.

The superior and inferior maxillary trunks of the fifth pair of cranial nerves will then be seen issuing from the skull; the former by the foramen rotundum, and crossing the spheno-maxillary fossa to the canal in the floor of the orbit (p. 35); the latter by the foramen ovale, and dividing into two branches; the anterior giving off five muscular branches to the masseter, buccinator, temporal, and external and internal pterygoid muscles; the posterior dividing into the auriculo-temporal, inferior dental, and gustatory branches (p. 35). A branch from the internal maxillary artery accompanies each of the nerves sent to the above-named muscles. The *otic*, or *Arnold's ganglion*, should be sought for, resting upon the inner surface of the inferior maxillary nerve, just below the foramen ovale.

The *auriculo-temporal nerve*, the terminal branches of which have been already dissected (p. 24), separates from the inferior maxillary near the skull; it passes outward beneath the external pterygoid muscle to the inner side of the articulation of the jaw, from which point it ascends with the temporal artery to ramify externally upon the side of the head.

The *inferior dental nerve* is the largest of the three branches into which the posterior trunk of the inferior maxillary divides; it lies at first beneath the external pterygoid muscle; afterward upon the internal pterygoid; it then enters the inferior maxillary bone at the inferior dental foramen, to emerge at the mental foramen (p. 22), and supply the lower lip and chin; this nerve is accompanied by the inferior dental branch of the internal maxillary artery. Before entering the canal in the bone, the inferior dental nerve gives off the *mylo-hyoid* branch, which, accompanied by the *mylo-hyoid artery*, a twig from the inferior dental, is continued along the inner aspect of the jaw to the mylo-hyoid muscle.

The *gustatory nerve* descends between the two pterygoid muscles to the side of the tongue, to the mucous membrane of which it is distributed; its final distribution will be described with the dissection of the tongue. The small *chorda tympani* branch of the facial nerve, arising within the temporal bone, and emerging from an aperture near the fissura Glaseri, joins the gustatory nerve, and, passing down in close apposition with it, establishes a connection with the submaxillary ganglion.

The *internal maxillary artery*, one of the terminal branches of the external carotid, passes inward behind the neck of the lower jaw, over the external pterygoid muscle, to the spheno-maxillary fossa; it is very tortuous, and sends off a large number of small branches; these are the

Masseteric,	Inferior dental,	Infra-orbital,
Buccal,	Tympanic,	Pterygo-palatine,
Temporal (deep),	Meningea media,	Spheno-palatine,
External pterygoid,	Meningea parva,	Posterior palatine,
Internal pterygoid,	Superior dental,	Vidian.

The five muscular branches, the first named of the above list, and the inferior dental, have been seen in connection with corresponding branches of the anterior trunk of the inferior maxillary and the inferior dental nerves. These branches are all capable of demonstration.

The *meningea media branch* ascends beneath the external pterygoid muscle, and enters the skull through the foramen spinosum of the sphenoid bone, to be distributed to the middle fossa of the cranial cavity.

The *meningea parva branch* passes through the foramen ovale, the same at which the inferior maxillary nerve emerges, and is also distributed to the middle fossa of the skull.

The *superior dental branch*, deeper seated than those already described, descends in a tortuous manner upon the tuberosity of the superior maxillary bone, sending branches through small foramina to the teeth, the antrum, and the mucous membrane of the gums.

The *infra-orbital branch* is the terminal portion of the internal maxillary artery; from the spheno-maxillary fossa it enters the infra-orbital canal with the superior maxillary nerve, and emerges on the face at the infra-orbital foramen (p. 19).

The remaining branches of the internal maxillary artery are of very small size, and, like the branches of the superior maxillary nerve, which they accompany, are incapable of demonstration, except by a special dissection.

The *internal maxillary vein* receives the veins corresponding to the branches of the artery, and unites with the temporal vein in a trunk, called the temporo-maxillary, which is one of the principal in forming the external jugular vein; the veins of the region form a plexus between the two pterygoid muscles.

The STYLO-GLOSSUS MUSCLE will be found, as its name implies, arising from the apex of the styloid process of the temporal bone; it is crossed by the gustatory nerve, and is inserted into the side of the tongue.

The STYLO-PHARYNGEUS MUSCLE lies below the preceding; it arises from the base of the styloid process, and is inserted into the pharynx and upper border of the thyroid cartilage; it passes between the external and internal carotid arteries, and the glosso-pharyngeal nerve turns over the lower part of its fleshy belly.

Between the stylo-glossus and stylo-pharyngeus muscles is a fibrous band, connecting the tip of the styloid process with the lesser cornu of the os hyoides; it is called the *stylo-hyoid ligament*, and is sometimes cartilaginous, or even osseous, in a part or the whole of its extent.

The styloid process is now to be cut through at its base, and with the two muscles just described turned downward; the fascia surrounding the internal carotid artery and the internal jugular vein is to be removed without disturbing the nerves in the vicinity of those vessels; they are the pneumogastric, glosso-pharyngeal, hypoglossal, sympathetic, and spinal accessory.

DEEP CERVICAL REGION.

THE INTERNAL JUGULAR VEIN is in close connection with, and lies upon the outer side of, the internal carotid artery; it emerges from the skull at the foramen jugulare, as the continuation of the lateral and petrosal sinuses; at about the level of the os hyoides its size is considerably increased by the junction with it of the facial, lingual, thyroid, and occipital veins. Between the skull and hyoid bone, this vein is sometimes called the *internal cephalic*, and below that point the internal jugular. Below the hyoid bone, it passes downward, parallel with and to the outer side of the common carotid artery, to join the subclavian vein, and with that forms the vena innominata. At the lower part of the neck its position corresponds to the interval between the sternal and clavicular attachments of the sterno-mastoid muscle.

The PNEUMOGASTRIC NERVE emerges from the jugular foramen, and will be recognized by the ganglion (ganglion of the trunk) peculiar to it; this is nearly an inch in length, and is surrounded by small nerves. The nerve lies between the jugular vein and internal carotid artery, and communicates with the hypoglossal, spinal accessory, and sympathetic nerves; it distributes branches to the parts about it, giving off a *pharyngeal* branch which unites with the glosso-pharyngeal nerve, and with which it forms a plexus on the pharynx; the *superior laryngeal* branch, of considerable size, passes inside the internal carotid artery to the larynx, which it enters between the hyoid bone and thyroid cartilage, perforating the thyro-hyoid membrane, and supplies the crico-thyroid muscle; it also gives off *cardiac* branches, which unite with those of the sympathetic.

The GLOSSO-PHARYNGEAL NERVE also escapes from the jugular foramen, and should be traced up to its ganglion, the *ganglion petrosum* or ganglion of Andersch, which lies close to the bone, and from which emanate the branches which unite it with the other nerves of this region. This nerve crosses over the internal carotid to the lower border of the stylo-pharyngeus muscle; it there assumes an almost transverse direction to the pharynx, and finally passes under the hyo-glossus muscle, to be distributed to the pharynx, tongue, and tonsil.

The SPINAL ACCESSORY NERVE, blending with the trunk of the pneumogastric, issues from the foramen jugulare, and is connected by small branches with the other nerves of this region. It passes outward, either over or under the jugular vein, to perforate the sterno-mastoid muscle at its upper part, and, uniting with the anterior cervical plexus, is distributed to the trapezius muscle (p. 41).

The INTERNAL CAROTID ARTERY ascends vertically from the upper border of the thyroid cartilage to the base of the skull; the rectus capitis anticus major muscle separates it from the vertebræ; except at its commencement, where it has a fusiform dilatation, it maintains the same size throughout, and gives off no branches; it enters the carotid canal of the temporal bone to emerge in the cranial cavity for the supply of the orbit and the encephalon. This vessel is sometimes tortuous instead of straight.

The *ascending pharyngeal artery*, seen in this dissection, is a branch of the external carotid, given off just as the common carotid bifurcates. It ascends, between the internal carotid and the pharynx, upon the spinal column; it gives a branch to the pharynx, and near the skull, where it becomes tortuous, it sends a branch through the foramen lacerum posterius to supply the posterior fossa of the skull (p. 45).

The HYPOGLOSSAL NERVE lies deep beneath the internal carotid; it issues from the skull at the anterior condyloid foramen, passes between the vein and artery, and descending curves round the occipital artery, and becomes superficial at the lower border of the posterior belly of the digastricus muscle, from whence, passing between the mylo-hyoid and hyo-glossus muscles, it is directed forward to the tongue and its muscles.

The SYMPATHETIC NERVE in the neck consists of a gangliated cord, which lies close to the spinal column; it is continuous with a similar gangliated cord in the thorax, and with several ganglionic bodies in the head connected with the three trunks of the fifth nerve. The cervical portion of the sympathetic has three ganglia; the *superior* one is fusiform in shape, and an inch or more in length; it is placed on the rectus capitis anticus major muscle, beneath the internal carotid artery and the trunks of the eighth nerve, which must be raised to expose it. It is connected with the other nerves, spinal as well as cranial, by minute filaments; some of these, called *nervi molles*, ramify upon the branches of the carotid artery, and form plexuses on their subdivisions. It sends a branch downward, behind the sheath of the carotid, to the heart, called the *superior cardiac nerve*.

The *middle cervical ganglion* is situated opposite the fifth cervical vertebra, lying upon or near the inferior thyroid artery, whence it is sometimes called the *thyroid ganglion;* it is of a rounded shape, and gives off branches which connect with the spinal nerves, and ramify upon the thyroid artery; it also gives off the *middle cardiac nerve*, which descends to the thorax, crossing the subclavian artery, and terminates in the cardiac plexus. The middle ganglion is sometimes wanting.

The *inferior cervical ganglion* occupies the interval between the first rib and the transverse process of the seventh cervical vertebra, and will be seen in a later stage of the dissection. (p. 59.)

DISSECTION VIII.

The upper portion of the sternum should now be removed, with the sternal half of the clavicle attached, and the ribs cut away on each side as far as the insertions of the scaleni muscles; this will expose the deep vessels of the neck in their relation to the heart; the cellular tissue which covers them is to be carefully removed by following the vessels downward from the point at which they are already dissected in the neck. The sterno-clavicular articulation should, however, be first examined, and this may be done before it is detached from the thorax.

STERNO-CLAVICULAR ARTICULATION.

The internal extremity of the clavicle is strongly connected with the upper bone of the sternum by several ligaments.

The *anterior sterno-clavicular ligament* is a stout band of fibres, passing from the anterior surface of the sternal end of the clavicle, downward and inward, to the upper and anterior part of the first bone of the sternum.

The *posterior sterno-clavicular ligament*, less developed than the preceding, occupies a situation, and has a similar origin and insertion on the inner aspect of the articulation.

The *inter-clavicular ligament* is a strong rounded cord, intervening between the superior surfaces of the clavicles, and is closely attached to the incisura semilunaris of the sternum.

The *costo-clavicular ligament* is a band passing obliquely forward from the under surface of the clavicle at its sternal end, to the cartilage of the first rib; sometimes the clavicle touches the rib, and has an articular surface at this point; the subclavius muscle lies in front of this ligament.

An *inter-articular fibro-cartilage* will be found on dividing these ligaments and detaching the clavicle; it is nearly circular in form, and thicker in the centre than at its circumference; it is adherent by its edges to the ligaments which surround the articulation. Its sternal surface is much more convex than its clavicular, which is nearly flat, and corresponds to a concavity in the sternum, so that the articulation in reality is between the sternum and the fibro-cartilage, and not between the sternum and clavicle.

The inter-articular cartilage divides the joint into two separate parts, each provided with a separate synovial membrane.

BASE OF THE NECK.

At the root of the neck, the region exposed behind the articulation above described, will exhibit the great venous branches which return the blood from the head and upper extremity.

The *internal jugular vein* descends the neck on the outer side of the carotid artery, and enters the subclavian vein, which is the continuation of the axillary vein; the internal jugular vein at this point is provided with two valves.

The SUBCLAVIAN VEIN lies in front of the artery of that name, and the union of this with the preceding vein forms the VENA INNOMINATA. The vena innominata is further reinforced by the *vertebral vein*, which descends at the side of the vertebral artery through the foramina of the transverse processes of the cervical vertebræ, and by the *inferior thyroid veins*, coming from the thyroid body and its neighborhood. The two venæ innominatæ unite to form the SUPERIOR VENA CAVA, and, owing to the destination of that vessel to the right auricle, the vena innominata of the left side is longer than that of the right, and its direction more nearly transverse; it lies upon the three primary branches of the aorta, and upon the upper part of the arch itself. The *thoracic duct* enters the left vena innominata at its commencement; the precise point may be demonstrated by inflating the duct in the thorax with a blow-pipe. The *ductus lymphaticus dexter*, which is the termination of the lymphatic vessels of the right side of the head and neck, right arm, and right side of the thorax, enters the right vena innominata near its commencement. The orifices of both these lymphatic ducts are provided with sigmoid valves to prevent admission of blood.

Between the subclavian vein and artery may be traced the pneumogastric and phrenic nerves.

The *pneumogastric nerve* has been already seen higher in the neck, lying between the carotid artery and internal jugular vein (p. 52); its lower part is now to be observed.

It follows a course a little different on one side from that on the other; on the right side the nerve passes in front of the subclavian artery, between it and the vein; on the left side it passes between the left subclavian and common carotid, and then crosses the arch of the aorta; below these points the two nerves resemble each other in their course, which is behind the root of the lung, along the œsophagus, to the stomach. As the right pneumogastric crosses the subclavian, it gives off the *recurrent laryngeal* branch which curves around that vessel, and is reflected upward to the larynx; on the left side this branch is given off as the nerve crosses the aorta, and it curves around its arch at the point where the ductus arteriosus is obliterated, to ascend like its fellow. In their course upward the recurrent branches lie on the inner side of the carotids, between the œsophagus and the trachea, to both of which they give branches; they terminate by filaments distributed to all the muscles of the larynx except the crico-thyroid, which is supplied by the superior laryngeal nerve.

The Phrenic Nerve is a medium-sized nerve arising from the anterior trunks of the third and fourth cervical nerves, and occasionally from the fifth; it descends obliquely upon the anterior scalenus muscle, crossing from its outer to its inner edge, and enters the thorax between the subclavian artery and vein, passing in front of the root of the lung, upon the pericardium, to the diaphragm, to which it is distributed.

The Arch of the Aorta gives rise to three vessels, collectively known as the brachio-cephalic trunks; the innominata, the left carotid, and the left subclavian; the number is sometimes increased by the left vertebral, which in a certain number of instances arises from the aorta, instead of from the left subclavian; when this irregularity takes place the supernumerary vessel is usually given off between the carotid and subclavian arteries. Occasionally, the number of primary branches is reduced to two. A very slight interval separates the aortic trunks from each other.

The Innominata Artery is a short trunk, three quarters of an inch in length; it lies upon the trachea, which it crosses somewhat obliquely, and then divides into the right carotid and right subclavian arteries. The innominata occasionally gives off a branch called the *middle thyroid*, which ascends tortuously in front of the trachea to the thyroid body; the existence of this anomalous branch, it will at once be seen, might essentially complicate the operation of tracheotomy.

The COMMON CAROTID ARTERIES will be seen to have a different origin, the right being from the innominata and the left directly from the aorta; the right is consequently shorter than the left by the length of the innominata. Separated at first only by the width of the trachea, they diverge as they ascend, their relations, except at their origin, being the same on the two sides (p. 42). The right carotid occasionally arises directly from the aorta.

The SUBCLAVIAN ARTERY differs, not only in origin, but in direction, on the two sides of the body, the right coming from the innominata and the left from the aorta direct; the right, moreover, is shorter than the left by the length of the innominata. The left subclavian ascends horizontally, and then turning suddenly, forms a right or an obtuse angle, while the right describes a gradual and regular curve. On both sides the artery passes beneath the scalenus anticus muscle, and rests upon the first rib, which is slightly grooved at the point where it reposes. The scalenus muscle separates the artery from its vein, which lies in front of the muscle; behind the artery are the scalenus posticus muscle, and the branches of the brachial plexus of nerves, which in a measure surround it. At the lower border of the first rib the subclavian becomes the axillary artery. The right subclavian has been observed to spring from the aorta direct, and is then most frequently the last in order of the primary aortic trunks; in such a case it crosses the neck obliquely, in front of the vertebral column and beneath the œsophagus, to regain its usual position. When this anomaly exists the carotid arteries not unfrequently spring from a common trunk, very short, and showing some marks of a tendency to divide. The left subclavian may arise in common with the left carotid.

The student should explore with his finger the tubercle on the first rib into which the scalenus anticus muscle is inserted, this being the guide to the position of the subclavian artery, when the surgeon desires to place a ligature upon it. He should also note the relations of the artery to the subclavian vein, the brachial plexus of nerves, the scalenus muscle, the clavicle, and the rib.

At this period of the dissection may be observed the variable height to which the pleural cavity sometimes extends above the first rib; a prolongation of two or even three inches has been observed. It will be seen that when such is the case, the thoracic cavity might easily be opened in the course of any operation in this locality.

The subclavian artery gives origin to the following branches, viz:—

  Vertebral,   Internal Mammary,
  Thyroid Axis,   Superior Intercostal.

The *vertebral artery*, the first and largest of the branches, ascends the neck upon the transverse processes of the cervical vertebræ until it reaches the sixth, when it enters the vertebral foramen and follows the canal formed by these foramina as far as the atlas; having passed through the foramen of the atlas, it curves around its articulating process, perforates the posterior occipito-atloid ligament, as well as the dura mater, and enters the foramen magnum to unite with the vertebral of the other side in forming the basilar artery. In the canal it lies in front of the anterior trunks of the cervical nerves, except the first and second, the former of which it crosses on its inner, the latter on its outer side. In its course the vertebral artery gives off branches to the membranes of the spinal cord and the deep muscles of the cervical region, and a *posterior meningeal* branch to the posterior fossa of the cranial cavity and falx cerebelli. Occasionally this artery does not enter the vertebral foramen till it has reached a point higher than the sixth vertebra. The arteries of the two sides may vary in size, in which case it is ordinarily the left which is largest; one of them may also, as has been stated, arise from the aorta; this irregularity has rarely been noticed except upon the left side.

The *thyroid axis* springs from the subclavian, close to the inner side of the scalenus anticus muscle; it does not always exist, as the branches to which it gives rise are not unfrequently given off directly from the subclavian; they are the—

    Inferior thyroid,
    Supra-scapular,
    Transversalis cervicis.

The *inferior thyroid artery*, a vessel of considerable size, passes obliquely inward, behind the common carotid, to the thyroid body, which it penetrates at its posterior surface, and in the substance of which it anastomoses with its fellow of the opposite side, and with the superior thyroid, a descending branch from the external carotid; in its course it supplies a branch called the *anterior cervical* to the posterior aspect of the trachea. The place of this artery is sometimes taken by a trunk called the *middle* or *lowest thyroid*, or *thyroid of Neubauer*, and which, springing from the innominata or the aorta, ascends in front of the trachea to the lower part of the thyroid body. This branch may be present also without the absence of the inferior thyroid.

The *supra-scapular artery* is a large branch which passes outward, behind the clavicle, to the scapula, where, crossing the supra-scapular notch, it is distributed to the dorsum of the scapula, as is described in Part Second, Dissection VI.

The *transversalis cervicis artery*, also a large branch, crosses the neck higher than the preceding, and divides into two branches, the *superfi-*

cialis cervicis, distributed to the under surface of the trapezius muscle, and the *posterior scapular*, which passes beneath the levator anguli scapulæ muscle and turns downward along the base of the scapula, as is described in Part Second, Dissection V. The posterior scapular sometimes arises as a separate trunk from the subclavian, outside of the scalenus anticus muscle; in this case the transversalis cervicis is of small size.

The *superior intercostal artery* is a descending branch of the subclavian, and can only be seen satisfactorily by cutting away the ribs in front; it turns over the neck of the first rib and supplies two, and occasionally three of the upper intercostal spaces.

The *profunda cervicis artery* usually arises in common with the preceding branch, although it may arise independently from the subclavian; it passes between the transverse processes of the seventh cervical and first dorsal vertebræ, and ascends the neck between the complexus and semi-spinalis colli muscles, to anastomose with a descending branch of the occipital artery, called the princeps cervicis. (See Part Second, Dissection V.)

The *internal mammary artery* is a large descending branch given off from the subclavian, and although sometimes arising from the thyroid axis, or in common with the subscapular, is very constant in its mode of origin; it is distributed to the inner surface of the anterior portion of the thorax, and is described in Part Second, Dissection III.

The *inferior cervical ganglion*, the third of the cervical ganglia of the sympathetic nerve (p. 54), lies behind the superior intercostal artery in the interval between the first rib and the transverse process of the seventh cervical vertebra. It is the largest of the three and irregularly rounded in shape; it is united with the middle cervical ganglion and with the chain of ganglia in the thorax. It gives off the *inferior cardiac nerve*, which, after joining the recurrent laryngeal branch of the pneumogastric nerve, is continued to the deep cardiac plexus lying upon the trachea, below the arch of the aorta.

The BRACHIAL PLEXUS can be better examined as to its origin and plan of formation at this period of the dissection than in connection with the axilla. It is formed by the union of the anterior branches of the last four cervical and first dorsal nerves; the fifth and sixth nerves emerging from the inter-vertebral foramina descend obliquely to unite in a trunk which speedily bifurcates; the eighth cervical and first dorsal ascend obliquely to unite in a large trunk which also bifurcates; the seventh continues by itself as far as the first rib, and then bifurcates, one branch uniting with the lower bifurcation resulting from the union of the fifth and sixth nerves, and its other branch uniting with the upper bifurcation resulting from the union of the eighth cervical and first dorsal nerves. We thus have four branches representing five original trunks.

DISSECTION IX.

The section known in the French amphitheatres as the *coupe du pharynx* is now to be made. The trachea and œsophagus, being divided opposite the top of the sternum, are to be turned up over the face; the chisel is to be applied at the apex of the angle formed by the upper part of these and the vertebral column, and, keeping it close to the latter, the basilar process is to be broken through behind the sella turcica: the skull is then to be separated into two halves, one anterior and the other posterior, by a section with the saw carried behind the styloid process and directed inward and a little forward; the two sides should be sawed separately and the two incisions should meet where the basilar process has been cut through with the chisel; the occiput and vertebral column will then constitute one of the pieces, while the other will be made up of the face, tongue, trachea, and pharynx.

Upon the latter of these pieces, the student should examine the lips, to see the *frena* formed by the mucous membrane and which attaches them to the lower and upper jaws upon the median line. The orifice of Steno's duct, in the cheek (p. 24), near the second molar tooth of the upper jaw, should also be sought for and explored by probing.

PHARYNX.

The lower jaw should be removed, except about half an inch on each side of the symphysis. The pharynx should be stuffed with tow, cotton-wool, or like material; and, when thus distended, the constrictor muscles will be easily dissected.

The PHARYNX is the dilated upper extremity of the alimentary tube, extending from the cavity of the mouth to the œsophagus, about four and a half inches; its inner surface is lined with mucous membrane, continuous with that of the mouth and œsophagus, and its walls are formed by muscles called constrictors.

The constrictor muscles are three in number on each side, and are so arranged that the lower overlaps the middle, and the middle the upper muscle; they are attached superiorly by an aponeurotic expansion to the base of the skull, and anteriorly to the larynx, hyoid bone, tongue, and bones of the nasal cavity.

The CONSTRICTOR INFERIOR is the most superficial of the three muscles; it arises from the side of the cricoid cartilage, and from the oblique line and upper and lower borders of the ala of the thyroid cartilage; from this origin the fibres are directed backward in a radiated manner to meet those of the corresponding muscle of the opposite side. They unite and form a tendinous raphé along the posterior median line of the pharynx. The lower fibres are continuous with the muscular fibres of the œsophagus. The recurrent laryngeal nerve passes under its lower border.

The CONSTRICTOR MEDIUS arises from the greater cornu of the os hyoides, and from the stylo-hyoid ligament; its fibres radiate from this origin, and blend along the middle line with the muscle of the other side, ending at its upper border in the aponeurosis of the pharynx. The glossopharyngeal nerve passes under the upper border of this muscle, and the stylo-pharyngeus muscle separates it from the superior constrictor.

The *stylo-pharyngeus muscle*, already described (p. 51), is again seen in this connection; arising from the base of the styloid process, and descending between the superior and middle constrictor, it is inserted partly into the pharynx, and partly into the upper border of the thyroid cartilage.

The *stylo-hyoid ligament*, already described (p. 51), is again seen extending from the tip of the styloid process of the temporal bone to the lesser cornu of the os hyoides.

The CONSTRICTOR SUPERIOR arises from the inner surface of the internal pterygoid plate, from the pterygo-maxillary ligament, the mucous membrane of the side of the mouth, and from the posterior part of the mylo-hyoid ridge of the lower jaw; from this extensive and irregular origin its fibres pass backward, and are inserted into the pharyngeal aponeurosis. The portion arising from the mucous membrane of the mouth has been described as a separate muscle, called the *glosso-pharyngeus*.

The *pterygo-maxillary ligament* is the raphé of union which exists between the superior constrictor and the buccinator muscles, one extremity being attached to the hamular process of the internal pterygoid plate, and the other to the extremity of the molar ridge of the lower jaw.

The *pharyngeal aponeurosis* connects the muscular part of the pharynx with the base of the skull, being attached superiorly to the basilar process of the occipital bone and the petrous portion of the temporal bone, and inferiorly losing itself in the muscular strata of the pharynx.

The pharynx is now to be laid open along the posterior median line; it is also to be dissected away, for a short distance on each side, from its aponeurotic attachment to the occipital bone.

The pharynx terminates in the œsophagus by a constriction opposite the cricoid cartilage. The mucous membrane of the pharynx is thicker, redder, and more abundantly supplied with muciparous glands than that of the œsophagus.

The ŒSOPHAGUS lies upon the bodies of the vertebræ behind the trachea, and, with two or three slight lateral curvatures, extends downward about nine inches to terminate in the stomach. Its walls, like those of the alimentary canal elsewhere, are composed of three layers, viz., muscular, cellular, and mucous. The muscular coat consists of longitudinal and circular fibres; these are red and well marked near the pharynx, but pale and indistinct as they approach the stomach. At the upper part, the longitudinal fibres, which are external, are arranged in three fasciculi, one anterior, arising from the cricoid cartilage, and two lateral, continuous with the inferior constrictor muscle of the pharynx. The circular, or internal fibres, are also continuous with those of the inferior constrictor. The mucous coat is very movable, and when not distended is thrown into longitudinal folds, so that the sides of the tube come in contact with each other.

The section of the pharynx exposes the posterior apertures of the nasal cavity, divided by the septum nasi. On each side of this will be found the trumpet-shaped orifice of the *Eustachian tube;* if the mucous membrane is removed, it will be found that this tube is composed of cartilage, and is nearly an inch in length; it is attached superiorly to a groove between the sphenoid and the petrous portion of the temporal bone, through which it communicates with the cavity of the tympanum. A bundle of muscular fibres, extending from the lower border of the Eustachian tube to the palato-pharyngeus muscle, is called the *salpingo-pharyngeus muscle;* it does not always exist.

PALATINE REGION.

The ISTHMUS FAUCIUM is the constriction between the mouth and the upper part of the pharynx, formed above by the soft palate, and below by the tongue.

The Soft Palate, or Velum Pendulum Palati, is a dependent structure which intervenes between the mouth and the pharynx; it is continuous with the roof of the mouth, or *hard palate*, superiorly; inferiorly its border is free, and from its centre hangs a pendulous body called the *uvula*. Two folds spring from each side of the uvula, and are continued downward on the sides of the isthmus faucium; these are the *arches* or *pillars* of the palate, the *anterior* reaching the side of the tongue at its root, and the *posterior* being continuous with the side of the pharynx. Between these pillars lies the tonsil.

The Tonsil is a collection of mucous follicles, forming a body, rounded in shape, but variable in size; the apertures of the follicles are sometimes very apparent; it is placed just above the tongue, and corresponds with the angle of the jaw externally.

The muscles of the palate are difficult of dissection, on account of their small size, and their close connection with surrounding parts. The upper attachment of the pharynx, with the superior constrictor, must be removed; the mucous membrane of the posterior surface of the soft palate is to be raised with great care, so as not to injure the muscular or tendinous fibres which compose the greater part of its substance; it may be made tense for dissection by drawing down and fastening the uvula with a hook or pin.

The Levator Palati arises from the under surface of the apex of the petrous portion of the temporal bone, and from the cartilage of the Eustachian tube; it then passes down by the side of the posterior nares, and spreads out in the structure of the soft palate as far as the median line.

The Tensor Palati arises from the scaphoid fossa at the base of the pterygoid plate, and from the outer part of the Eustachian tube; it terminates in a tendon which winds round the hamular process of the pterygoid plate, and, expanding into an aponeurosis, is inserted into the posterior border of the horizontal portion of the palate bone; it also, with the muscle of the opposite side, helps to form the aponeurotic basis of the soft palate, lying beneath the tendon of the preceding muscle.

The Palato-pharyngeus Muscle forms the posterior pillar of the soft palate; it arises by a broad expansion from the lower part of the soft palate, and, joining with the stylo-pharyngeus muscle, is inserted into the posterior border of the thyroid cartilage.

The Azygos Uvulæ, situated upon the posterior surface

along the middle line of the soft palate, consists of two bundles of very pale fibres, arising from the spine at the posterior border of the hard palate, and terminating at the free extremity of the uvula.

The PALATO-GLOSSUS MUSCLE, or CONSTRICTOR ISTHMI FAUCIUM, forms the projection of the anterior pillar of the soft palate; it arises from the aponeurosis of the soft palate in front of the tensor palati, and is inserted into the lateral surface and dorsum of the tongue.

OTIC AND MECKEL'S GANGLIA.

The tongue, hyoid bone, and larynx, may now be separated from the portion of the skull to which they are attached, and reserved for further examination. The spheno-maxillary fossa is to be searched for certain ganglia of the sympathetic system.

The OTIC, or ARNOLD'S GANGLION, if not already examined in connection with the inferior maxillary nerve, will be found as a small ovoid body just below the foramen ovale, on the inner side of the inferior maxillary nerve, and at the point where it is joined by the second root of the trifacial nerve. This ganglion gives branches to the tensor tympani and tensor palati muscles, communicates with the nerves of its vicinity, and sends a branch to the Vidian nerve, called *nervus petrosus superficialis minor;* but its connections are by twigs so minute that their examination can only be accomplished by the aid of a lens and by a special dissection.

In the spheno-maxillary fossa, and connected by short branches with the superior maxillary nerve, is the *spheno-palatine,* or *Meckel's ganglion;* it lies below the superior maxillary nerve, above the posterior palatine canal, and outside the spheno-palatine foramen; its branches are directed upward to the orbit, downward to the mouth, inward to the nose, and backward to the pharynx and Vidian canal; the principal of these are as follows:—

The *large palatine* branch passes through the posterior palatine canal and forward along the roof of the mouth nearly to the incisor teeth, where it unites with the naso-palatine branch of the same ganglion. The *small palatine* branch descends to supply the soft palate, the tonsils, and the uvula.

The *naso-palatine* branch crosses the roof of the nasal fossa to the septum nasi, on the side of which it descends to the anterior palatine canal, where it unites with the large palatine branch.

The *Vidian nerve* passes through the Vidian or pterygoid canal, then enters the cranium through the cartilaginous tissue closing the foramen lacerum anterius, and, lying in a groove on the surface of the petrous portion of the temporal bone, takes the name of *nervus petrosus superficialis major*, and enters the hiatus Fallopii to join the gangliform enlargement of the facial nerve.

These branches can hardly be found except by a special dissection.

Before discarding the piece of bone which the student has just been examining, he should open the carotid canal to see the *temporal portion of the internal carotid artery*. The canal can be opened by bone forceps, or by a small chisel carefully handled. The artery at first ascends, is then directed forward horizontally, and lastly turns upward into the cranium. In the canal it is surrounded by branches from the superior cervical ganglion of the sympathetic nerve; it gives off a minute branch to the tympanum.

NASAL FOSSÆ.

To expose the nasal cavities, the saw should be placed at the side of the crista galli, and carried through the nasal and frontal bones to the roof of the mouth; this separates the portion of skull into two halves, leaving the septum nasi attached upon one side.

The two cavities on either side of the septum of the nose are called the *nasal fossæ;* they communicate with the pharynx and face by apertures, called anterior and posterior nares, and also with the sinuses of the frontal, ethmoid, sphenoid, and superior maxillary bones. The *septum* of the fossæ is partly osseous and partly cartilaginous, being formed by the perpendicular plate of the ethmoid bone, the vomer, and the triangular cartilage of the nose. The septum is usually more or less bent to one or the other side.

On the outer boundary of the nasal fossæ will be found the three convoluted *spongy or turbinated bones*. The two superior are processes of the ethmoid bone, and are known respectively as the *superior* and *middle turbinated bones;* the lower one is an independent bone, called the *inferior turbinated bone*. These bones overhang spaces called meatuses; they are three in number; the *superior meatus* is the smallest; it is overhung by the superior turbinated bone, and occupies nearly the posterior half of the outer aspect of the nasal fossa; the posterior ethmoidal sinuses open into it in front, and in the dried bone at its posterior part will be found the spheno-palatine foramen, by which the nerves and vessels enter the nose. The middle turbinated bone overhangs the *middle meatus*, which extends nearly the whole length of the outer wall of the fossa; it communicates anteriorly with the frontal sinuses and anterior ethmoidal cells, and near its middle is a small oblique aperture the size of a crow-quill, which leads to the cavity of the antrum. The *inferior meatus*, overhung by the inferior turbinated bone, also extends the whole length of the nasal fossa; at its anterior part is the opening of the *nasal duct*, and on a level with this, but more posteriorly, the Eustachian tube, which connects the cavity of the tympanum with the pharynx, may be seen.

The mucous membrane which lines the fossæ and meatuses is continuous with the skin anteriorly, and the pharyngeal mucous membrane posteriorly, and is called the *pituitary* or *Schneiderian membrane*.

DISSECTION X.

TONGUE.

The Tongue is made up of muscular fibres, interspersed with fat and cellular tissue, and enveloped by mucous membrane. It is separated by a fibrous septum into two lateral halves, and is attached by its base to the hyoid bone; anteriorly it is connected by a fold of mucous membrane, called the *frenum linguæ*, with the symphysis of the lower jaw, and posteriorly with the epiglottis, by a central and two lateral folds of mucous membrane, called collectively the *glosso-epiglottidean ligament;* the central fold is called the *frenum epiglottidis*. On each side of the frenum linguæ is a papilla, which marks the orifice of the duct of Wharton (p. 47). On each side of the frenum there is also a prominence caused by the *sublingual gland;* this is a flattened and elongated conglomerate gland, similar to the other salivary glands, but smaller; it pours its secretion into the mouth by numerous small orifices, called the *Rivinian ducts*, which open on each side of the frenum.

The upper surface of the tongue is covered with numerous papillæ, called from their shape *conical* and *filiform:* the latter are most numerous along its sides; scattered among these are a few more prominent papillæ, which from being larger at the top than at the base are hence called *fungiform*. At the base of the tongue, arranged across it in the shape of the letter V, are eight or ten papillæ called *caliciform*, or *circumvallatæ;* they are pedunculated, and surrounded by a fold of mucous membrane; at the apex of the two rows of these papillæ is a deep mucous follicle, called the *foramen cæcum*.

Some of the muscles of the tongue have been already described and divided; their lingual ending may be further followed out, and the intrinsic muscles dissected by removing the mucous membrane; to do this the tongue should be fixed, by large pins driven through it, to the table on which it lies.

The HYO-GLOSSUS MUSCLE arises from the greater and lesser cornua and body of the os hyoides, and is inserted beneath the stylo-glossus into the side of the tongue. That portion of the muscle which arises from the body of the hyoid bone and passes obliquely backward, has been called the *basio-glossus;* that from the great cornu passes forward, which has been called the *cerato-glossus;* while a few fibres from the lesser cornu, spreading along the side of the tongue, have been named the *chondro-glossus*. These names were applied to the different parts of the muscle by Albinus. The existence of the chondro-glossus is disputed.

Between the hyo-glossus and the genio-hyo-glossus muscles, and connected with that portion of the hyo-glossus described as the chondro-glossus, is a bundle of fibres running longitudinally from the base to the apex of the tongue; this is called the *inferior lingualis* muscle. The *superior lingualis* muscle is a thinner and less distinct band of fibres just beneath the mucous membrane upon the dorsum of the tongue, arising from the frenum epiglottidis, and extending forward to its apex.

The hyo-glossus must be divided at its hyoid attachment, and reflected, in order to expose the next muscle.

The GENIO-HYO-GLOSSUS MUSCLE is triangular in shape; it arises from the symphysis of the lower jaw, above the genio-hyoid muscle, and expanding fan-shaped, is inserted into the whole length of the tongue and into the body of the hyoid bone; by making a longitudinal section of the tongue in the median line, the direction of its fibres will be plainly visible.

A section thus made will also show the fibrous *septum* which separates the tongue into two halves, giving to it the character of duality which belongs to all the special senses; although rudimentary in the human species, this separation into two parts is marked in the tongues of certain birds and reptiles.

The *lingual artery* will be found resting upon the genio-hyo-glossus muscle; it gives a branch to the dorsum of the tongue, and is continued forward to its apex under the name of the *ranine artery*.

Upon the hyo-glossus muscle the *gustatory nerve* will be seen; this is a branch of the inferior maxillary trunk of the fifth nerve; it anastomoses with small branches of the hypoglossal, and is distributed to the mucous membrane and papillæ of the tongue.

The *hypoglossal nerve* also crosses the hyo-glossus muscle, and, after penetrating the genio-hyo-glossus, is continued forward to the apex of the tongue in the substance of that muscle. It supplies the muscular structure of the tongue.

The *glosso-pharyngeal nerve* passes under the posterior border of the hyo-glossus muscle, and supplies the dorsum and lateral mucous membrane of the tongue, the tonsils, and the fauces.

LARYNX.

The tongue should now be separated from the larynx without injuring the epiglottis; the hyoid bone should remain attached to the larynx. The larynx should be extended, and made fast to the table by means of pins; the pharynx and extrinsic muscles should be dissected away from its cartilages and from the hyoid bone.

The hyoid bone is connected with the thyroid cartilage of the larynx by a membrane called the *thyro-hyoid membrane*, which includes the entire space intervening between the upper border of the thyroid cartilage and the upper border of the hyoid bone; this membrane is perforated by the superior laryngeal nerve and the superior laryngeal artery.

The CRICO-THYROIDEUS MUSCLE is a triangular muscle, and the largest of the special muscles of the larynx; it arises from the front and lateral part of the cricoid cartilage, and passes outward and backward, to be inserted into the external and internal surfaces of the lower border of the thyroid cartilage.

The CRICO-ARYTENOIDEUS POSTICUS MUSCLE arises from the depression on the side of the vertical ridge of the cricoid cartilage posteriorly; its fibres converge and are inserted into the muscular process of the arytenoid cartilage.

The ARYTENOIDEUS MUSCLE covers in and unites the posterior surfaces of the two arytenoid cartilages; the fibres of this muscle consist of fasciculi, having partly a horizontal or transverse, and partly an oblique direction.

To expose the remaining muscles of the larynx, one side of the thyroid cartilage must be removed.

The CRICO-ARYTENOIDEUS LATERALIS MUSCLE arises from the lateral part of the upper border of the cricoid cartilage, and is inserted into the muscular process of the arytenoid cartilage.

The THYRO-ARYTENOIDEUS MUSCLE arises from the angle of the thyroid cartilage and is a complex muscle, imperfectly demonstrable in an ordinary dissection. One portion, quadrangular in shape, passes backward to be inserted into the vocal process of the arytenoid cartilage, and forms the lower surface of the vocal cord. A second portion also runs backward and is inserted into the lower part of the antero-external surface of the arytenoid cartilage, helping to form the lower wall of the ventricle of the larynx. A third portion sends its fibres in different directions; some pass backward to the lower part of the outer edge of the arytenoid cartilage; others passing along the outer wall of the ventricle of the larynx lose themselves in the ary-epiglottidean fold; others again pass partly into the ventricular cord and partly up to the epiglottis.

The cavity of the larynx is divided into two parts by a constriction caused by the prominence of two elastic fibrous ligaments, attached in front to the angle of the thyroid cartilage, and posteriorly to the bases of the arytenoid cartilages. These are called the *vocal cords*, and the space between them is called the *rima glottidis*. Immediately above and running parallel with the vocal cords are two folds of mucous membrane called the false cords, or, more properly, the *ventricular cords* or *bands*. The space between the vocal cords and the ventricular bands is called the *ventricle of the larynx*.

In the ventricle of the larynx the mucous membrane forms a deep pouch, best demonstrated by stuffing it with cotton-wool; this is called the *sacculus laryngis*, and is directed upward, sometimes as high as the upper border of the thyroid cartilage.

The lateral boundaries of the upper laryngeal aperture are formed by folds of mucous membrane, called the *ary-epiglottidean folds*, which pass from the sides of the epiglottis, backward, downward, and inward, in a semi-circular form, toward each other. In these folds, posteriorly, are found the cartilages of Wrisberg, and the cartilages of Santorini.

The larynx is supplied with nerves from the superior and inferior laryngeal branches of the pneumogastric nerve. The superior laryngeal penetrates the thyro-hyoid membrane, and supplies the mucous membrane and crico-thyroid muscle. The inferior or recurrent laryngeal nerve passes beneath the ala of the thyroid cartilage, and supplies all the special muscles of the larynx except the crico-thyroid.

These nerves are accompanied by the laryngeal branches of the superior and inferior thyroid arteries, which follow them in their distribution to the mucous membrane and muscles.

<small>The hyoid bone and the cartilages of the larynx are now to be denuded, care being taken to leave the ligaments which unite the different cartilages, as well as the membrane between the larynx and hyoid bone.</small>

The HYOID BONE consists of a body and four cornua, two greater and two lesser. The *greater cornu* extends backward, decreasing gradually in size, and ends posteriorly in a tubercle; the *lesser cornu* is directed upward, and is about the size and shape of a kernel of rice; it marks the union of the body of the bone with the greater cornu, and is seldom co-ossified with it; the stylo-hyoid ligament is attached to it. (p. 51.)

The THYROID CARTILAGE of the larynx consists of two quadrilateral halves united anteriorly at an acute angle, forming a projection called the *pomum Adami*. The upper border, above the pomum Adami has a deep notch called *incisura thyroideæ superior;* the lower border has a slighter notch called *incisura thyroideæ inferior*. Posteriorly the cartilage has a thick border, which terminates above and below in a cornu; the upper being the longest, and the lower the stoutest. The upper cornu is connected to the greater cornu of the hyoid bone by a ligament, called the *thyro-hyoid;* a sesamoid bone, or cartilage, is usually found in it. The lower cornu articulates with the cricoid cartilage. The inner surface of this cartilage is smooth, the outer is marked by an *oblique ridge*, extending from its upper cornu to the middle of the lower border.

This cartilage is connected to the cricoid cartilage by a membrane called the *crico-thyroid*.

The CRICOID CARTILAGE forms a complete ring, and is thicker than the thyroid; it is broader behind than in front, and resembles somewhat in its shape a signet ring; the inner surface is smooth, the outer is rough for the attachment of muscles; the lower border is straight, and united by a fibrous membrane with the first ring of the trachea; the upper border is irregular.

The ARYTENOID CARTILAGES are two in number, one on each side, and are placed on the upper border of the cricoid cartilage at the back of the larynx; they are pyramidal in shape, having a base and three surfaces; at the lower part there is externally a short process called the *muscular process*, and anteriorly a longer pointed process, the *vocal process*. Attached to the apex of each cartilage is a small fibrous body, called the *cartilage of Santorini*.

The CUNEIFORM CARTILAGES, or CARTILAGES OF WRISBERG, are attached to the middle of the external surface of the arytenoid cartilages; they are sometimes wanting.

The EPIGLOTTIS is composed of yellow elastic fibre, and resembles in shape a rounded leaf with the stalk downward; it is often indented in the middle of its free edge and rolled in from the sides; it is connected with the tongue by a central and two lateral folds of mucous membrane, known as the *glosso-epiglottidean ligament;* the central fold is called the frenum epiglottidis. A fibrous band, called the *thyro-epiglottic ligament*, unites the epiglottis to the notch in the anterior border of the thyroid cartilage below; it is also connected with the hyoid bone by strong bands of fibrous tissue called the *hyo-epiglottic ligament*. Between the epiglottis and the hyoid bone is a mass of yellowish fat, which has been improperly named the *epiglottidean gland*.

The TRACHEA is composed of from sixteen to twenty incomplete cartilaginous rings, each forming about three-fourths of a circle; some of them blend together, but for the most part they are united by a fibrous tissue, extending across between their ends and forming a flattened wall which completes the tube. The posterior wall of the trachea, in addition to this fibrous membrane, has a layer of muscular tissue closely connected with the mucous membrane, and its surface is dotted with well-marked muciparous glands. The continuation of the trachea is described in connection with the lungs.

DISSECTION XI.

PREVERTEBRAL REGION.

The dissection now reverts to that portion of the vertebral column put aside after the separation of the pharynx, viz., the prevertebral region. The removal of a small amount of cellular tissue exposes several muscles; the superior of which, the rectus capitis anticus major and minor, are usually found mutilated by the cutting through of the occipital bone.

The RECTUS CAPITIS ANTICUS MAJOR MUSCLE, superiorly the most superficial of the prevertebral muscles, arises by tendinous slips from the anterior tubercles of the transverse processes of the third, fourth, fifth, and sixth cervical vertebræ; these unite to form a single belly, and the muscle is inserted into the basilar process of the occipital bone.

The LONGUS COLLI MUSCLE, partly concealed by the preceding, is, however, superficial at its lower part, where it lies upon the bodies of the vertebræ. This muscle is pointed above and broad below, and consists of an external and an internal part, the former being vertical, and the latter oblique. The internal part arises by fleshy and tendinous processes from the bodies of the two upper dorsal and two lower cervical vertebræ; the external part arises from the anterior tubercles of the transverse processes of the third, fourth, fifth, and sixth cervical vertebræ; the two portions then blend together, and are inserted by four slips into the lower border of the bodies of the four upper cervical vertebræ; some of the most external fibres of the lower part of this muscle are occasionally attached by separate tendons to the transverse processes of the lower cervical vertebræ.

The RECTUS CAPITIS ANTICUS MINOR MUSCLE lies beneath the rectus anticus major, which must therefore be removed; it arises from the anterior border of the lateral half of the atlas, and is inserted into the basilar process of the occipital bone.

The RECTUS CAPITIS LATERALIS MUSCLE arises from the

transverse process of the atlas, and is inserted into the rough surface of the occipital bone external to its condyle.

The SCALENUS ANTICUS MUSCLE arises from the anterior tubercles of the transverse processes of the third, fourth, fifth, and sixth cervical vertebræ, and is inserted into a tubercle on the upper surface of the first rib; this tubercle is important to notice, since it may be felt by the finger in searching for the subclavian artery when that vessel is to be tied, and the artery lies directly behind this insertion of the muscle. The vertebral origin, it will be observed, corresponds with that of the rectus anticus major muscle, of which it may in some sort be considered the continuation.

The SCALENUS MEDIUS MUSCLE arises from the posterior tubercles of the transverse processes of all the cervical vertebræ, except the first, and is inserted by two fleshy bellies into the first and second ribs, the one inserted into the second being smaller than that into the first rib; the attachment to the second rib is often described as a separate muscle, under the name of *scalenus posticus*.

Between the anterior tubercles of the transverse processes of the cervical vertebræ, are the anterior pairs of the *inter-transversales muscles*, the first pair being between the atlas and axis, and the last between the last cervical and first dorsal vertebræ. The lower anterior inter-transversalis muscle is often wanting. Between the pairs, except in the first two spaces, the anterior primary branch of the cervical nerves makes its exit; beneath the posterior muscle the posterior primary branch of the same nerve emerges.

The anterior division of the *first spinal* or *sub-occipital nerve*, is given off from the common trunk while the latter lies upon the posterior arch of the atlas; it then curves downward in front of the transverse process of the atlas, and forms a loop of communication with an ascending branch of the second cervical nerve; it supplies the rectus lateralis and rectus anticus minor, and communicates with the pneumogastric, hypoglossal, and sympathetic nerves.

The anterior division of the *second cervical nerve* passes over the lamina of the axis, is directed forward, outside the vertebral artery, and beneath the inter-transversalis muscle of the first space, to join the cervical plexus.

The *vertebral artery*, already described at p. 58, may be further examined, by removing the inter-transversales muscles; it will then be seen lying in the canal of the transverse processes of all the cervical vertebræ, except the seventh; after emerging from the atlas, it winds round its articular process, and, piercing the posterior occipito-atloid ligament as well as the dura mater, enters the foramen magnum. Its small offsets to the spinal membranes can now be seen at their origin.

LIGAMENTS OF THE FIRST TWO VERTEBRÆ.

The cervical vertebræ are connected with each other by ligaments similar to those of the rest of the vertebræ, which are described in Part Second, Dissection X. The two upper vertebræ are, however, connected by special ligaments with each other, and with the occipital bone.

* The removal of all the muscles, and other extraneous tissues about the base of the skull and the upper part of the vertebral column, will be necessary to expose these articulations.

The OCCIPITO-ATLOID ARTICULATION is maintained by two ligaments. The *anterior occipito-atloid ligament* consists of a rounded cord attached above to the basilar process of the occipital bone, and below to the anterior tubercle of the atlas; beneath this is a broad membrane, extending from the anterior margin of the foramen magnum to the anterior arch of the atlas; this latter is sometimes described separately as the deep anterior ligament. The *posterior occipito-atloid ligament* extends from the posterior margin of the foramen magnum to the posterior arch of the atlas; it is perforated by the vertebral arteries and by the sub-occipital nerve. *Lateral ligaments*, consisting of stout fibrous bands, extend from the bases of the transverse processes of the atlas to the jugular processes of the occipital bone; *capsular ligaments* surround the articulation of the condyles of the occipital bone with the articulating surfaces of the atlas, and the articulation of the atlas and axis is similarly provided.

The ATLO-AXOIDEAN ARTICULATION is maintained externally by a thin membrane, which unites the bodies of the atlas and axis anteriorly, and their arches posteriorly.

To see the ligaments inside the spinal canal, the posterior arches of the atlas and axis must be cut away, as well as the occipital bone so far as it forms the posterior half of the foramen magnum; the dura mater is to be removed.

The ligaments constituting the OCCIPITO-AXOIDEAN ARTICULATION are peculiar in arrangement, and maintain the relation of the bones during rotatory and nodding movements.

The *occipito-axoid ligament*, or the *apparatus ligamentosus colli*, is attached above by a broad expansion to the basilar process of the occipital bone, in front of the foramen magnum; below it is continuous with the posterior common ligament of the vertebral column; its deeper fibres, however, have a special insertion into the posterior part of the body of the axis, and into the superior border of the transverse ligament.

The occipito-axoid ligament is to be removed in such a way as to leave the portion which is inserted into the transverse ligament.

The *odontoid ligaments* are two strong bands extending from the sides of the apex of the odontoid process to a depression on the inner surface of the condyles of the occipital bone. A third band extends from the tip of the odontoid process to the anterior edge of the foramen magnum; it bears the name of *ligamentum suspensorium*. As these ligaments control the movements of the cranium in rotation they are sometimes called the *check ligaments*.

The *transverse ligament* extends across the odontoid process from a tubercle on the inner surface of one articular process of the atlas to a similar point on the other; its upper border receives a part of the occipito-axoid ligament, and its inferior border sends some fibres downward to the base of the odontoid process; from this arrangement of its fibres it is sometimes called the *crucial ligament*.

The removal of the odontoid and transverse ligaments will show an articulating surface in front of the odontoid process, where it rests against the atlas, and another behind, where it comes in contact with the transverse ligament; these are both provided with synovial membranes.

DISSECTION XII.

ANATOMY OF THE EYE.

The structure of the eye is best studied upon that of some animal. The market-house will almost always furnish those of an ox, calf, or sheep. A number of them is desirable, and their dissection should be conducted in water. A soup-plate is a convenient dish for this purpose. The muscles and the mucous membrane should all be removed and the eyeball left with the optic nerve alone attached to it. In order to obtain a general idea of the different parts, the student will do well to make both a longitudinal and a transverse section of the eyeball, and this he will easily accomplish if he will first freeze the eye by immersing it in a mixture of ice and salt.

The ORGAN OF VISION is composed of certain parts essential to sight and of others requisite for their protection. Its sentient constituent is an expansion of the optic nerve (retina). A series of transparent parts (cornea, lens, vitreous humor) bring the rays of light to a focus upon the retina; and a movable curtain (iris) regulates the extent of their admission. To defend these structures, a dense tunic (sclerotic) is arranged around them, and to absorb the superabundant rays of light there is an internal tunic (choroid) provided with a layer of pigment.

The SCLEROTIC TUNIC extends from the entrance of the optic nerve to the cornea, and forms five-sixths of the surface of the eyeball. The optic nerve perforates it a little to the inner side of its centre; a marked constriction characterizes it just previous to its entrance; if the nerve be drawn out, it will be seen that it leaves an orifice in the sclerotic, tapering inward, and which at the bottom is perforated by minute holes, through which the fibres of the nerve penetrated; this spot has hence been called the *lamina cribrosa*. The inner aspect of the sclerotic is flocculent with the ends of ruptured vessels and nerves; externally it is smooth except where muscles are attached, and it is thickest at the posterior part.

The CORNEA is a clear and diaphanous structure which forms the anterior wall of the anterior chamber of the eye and admits the rays of light. It is circular in form, and, like the crystal of a watch, is convex anteriorly and concave posteriorly; at its circumference it blends with the sclerotic by continuity of tissue; its structure is laminated, and it is covered superficially by the conjunctiva (p. 17); its internal surface is lined by a structureless membrane, called the *membrane of Descemet* or *of Demours*. In its healthy state the cornea is devoid of vessels; after death it becomes flaccid and opaque from infiltration of the aqueous humor.

The *vascular coat of the eyeball* is internal to the sclerotic; it is a thin membrane made up of bloodvessels and pigment cells, divisible into three parts: a posterior portion, corresponding to the sclerotic, and which is called the choroid; an anterior portion opposite the cornea, called the iris; and an intermediate ring on a level with the union of the sclerotic and cornea, consisting of the ciliary muscle and processes.

<small>To show the choroid, the sclerotic is to be cut through just behind the cornea, and from this circular incision three or four slits are to be made with the scissors toward the optic nerve; the resulting flaps are then to be carefully reflected. In another eye the cornea should be removed and the iris gently torn away with the forceps; this will expose the ciliary processes.</small>

The CHOROID is a thin pigmentary membrane, extending from the optic nerve, which perforates it, to the anterior part of the eyeball, as far forward as a white ring at the line of union between the sclerotic and cornea, which is the ciliary muscle; at this point the choroid bends inward behind the iris to end in a series of plaited folds, called the *ciliary processes*, arranged around the lens in the form of a circle. The pigment of the choroid is principally contained in an internal epithelial layer, and is easily detached and washed away. In many animals, at the bottom of the eye, there is a brilliant arrangement of fibres on the internal surface of the choroid, which shines with a metallic lustre, and is called the *tapetum oculi*.

The CILIARY LIGAMENT, or, as it is now more properly termed, the CILIARY MUSCLE, will be seen as a white ring sometimes called the *annulus albidus*, situated outside that part of the choroid which turns inward to form the ciliary processes, and opposite the junction of the cornea and sclerotic coats. The ciliary muscle constitutes the most anterior part of the choroid; its muscular character is not capable of demonstration in an ordinary dissection, but it fulfils an important office in connection with the sense of vision. Between the ciliary muscle and the sclerotic is a minute vascular canal called the *canal of Schlemm*, sometimes also *of Fontana*.

The IRIS is a vascular and muscular structure, sufficiently displayed in the dissections already made; it is circular in form, and is suspended vertically in front of the crystalline lens; its anterior surface is free in the aqueous humor, and marked by lines converging toward the pupil; its posterior surface is covered by a thick layer of pigment, to which the name *uvea* has been applied, and is in contact with the lens; its circumference corresponds to the point at which the sclerotic and cornea blend, and it is connected with the choroid by means of the ciliary muscle. The muscular fibres of the iris are unstriated, and are both circular and radiating.

The CHAMBERS OF THE AQUEOUS HUMOR are the spaces between the lens behind and the cornea in front, separated by the iris into two compartments, *anterior* and *posterior*. The anterior chamber is the larger of the two; their boundaries will be best seen in an eye that is bisected longitudinally. The *aqueous humor* is a pure transparent fluid; it disappears by transudation a short time after death.

The NERVOUS COAT, or the RETINA, requires to be dissected in a recent eye; it is the most internal of the three coats of the eyeball and one of the most delicate of its structures; it lies between the choroid and the vitreous humor, upon which it is moulded.

The retina may be seen by gently tearing away the choroid in the eye that was used for the examination of that membrane, or by emptying an eye of its vitreous humor after the removal of an anterior segment.

The retina expands as far forward as the outer edge of the ciliary processes, where it forms a scolloped border, called the *ora serrata*, and may be traced backward to the lamina cribrosa, the point at which the optic nerve enters the eyeball; a thin hyaloid membrane, called the *zonula ciliaris*, or *zone of Zinn*, is attached to the anterior border of the retina, and continued to the anterior surface of the lens. This membrane, which is composed of folds into which the ciliary processes fit, becomes stained with their pigment, so as to leave a series of radiating lines around the lens. On looking through the vitreous body, the central artery of the retina may be seen entering by a central foramen in the lamina cribrosa called the *porus opticus*. A circular spot at the posterior part of the retina is called the *foramen of Soemmering;* and this is surrounded by a yellow halo called the *limbus luteus*.

The CRYSTALLINE LENS lies in a depression in the front of the vitreous humor, and behind the pupil of the iris. It has a proper capsule of transparent membrane, and is held in its place by a suspensory ligament, which is formed by the zonula ciliaris in front and by a similar hyaloid membrane posteriorly, the two uniting behind the ciliary processes; the space between these two layers is called the *canal of Petit;* this may be demonstrated by the blow-pipe, and the inflation of the folds in the anterior membrane gives the canal a beaded appearance when distended. The lens itself is a transparent double convex body, the posterior convexity being greater than the anterior: its consistency is greater at the centre or *nucleus* than it is externally. By allowing it to lie in water for a little while, each surface will separate into three parts by lines radiating from the centre, those of one side being intermediate in position to those of the other. If the lens be hardened either in alcohol or by boiling, it will be easy to separate it into laminæ, like those of an onion, each lamina being divided into three segments by the diverging lines, just spoken of.

The VITREOUS HUMOR is a jelly-like, transparent substance, filling the greater part of the eyeball behind the iris; it is contained in a thin and delicate membrane called the *hyaloid*, which not only envelops it externally, but penetrates into its interior, and, forming a sort of trabeculated arrangement, helps to support it and retain it in place.

DISSECTION XIII.

ANATOMY OF THE ENCEPHALON.

The density of the cerebral substance increasing with age, the brains of persons advanced in life are the most desirable for examination. The autopsy-room is the only place where they can be obtained fresh enough for dissection, and the firmness of their tissue may be improved by several days' immersion in alcohol. For inspection, the brain is to be placed with its base uppermost, and with its convexity resting in a coil of cloth, to maintain it in position.

MEMBRANES AND VESSELS.

The dura mater has been previously described (Dissection IV.); the arachnoid and pia mater are still connected with the brain after its removal; all of these membranes are prolonged into the vertebral canal, and can be seen in dissecting the spinal cord.

The ARACHNOID, a thin serous membrane, the parietal part of which has been already observed lining the inner surface of the dura mater, is reflected over the pia mater, and may be demonstrated at almost any point where it passes across the intervals between one convolution of the brain and another. It covers the encephalon loosely, and leaves a considerable space between it and the convolutions; this space, called the *sub-arachnoid*, is greatest at the base of the brain, and here as elsewhere contains more or less fluid, which, from the continuity existing between the sub-arachnoid space of the cranium and that of the spinal canal, is called the *cerebro-spinal fluid*. The cerebro-spinal fluid is, however, often lost in the course of dissection, from wounding a dependent portion of the membrane, through which it drains off from both the cerebral and spinal spaces.

The PIA MATER is the vascular covering of the brain, and this, likewise, envelops both the encephalon and spinal cord; it is in close contact with the cerebral substance, and dips into the fissures and convolutions; it also penetrates into the interior of the brain, to supply vessels to the walls of the cavities there existing.

In connection with the membranes will be found the arteries of the brain, derived from the vertebral and internal carotid arteries.

The vertebral artery (p. 58) supplies the cerebellum and posterior lobes of the cerebrum; it ascends by the side of the medulla oblongata, and at the posterior border of the pons Varolii unites with its fellow to form the basilar artery. Prior to their union they give off certain small branches distributed to the spinal cord and its membranes, as well as the posterior meningeal artery to the falx cerebelli and cerebellar fossæ.

The *inferior cerebellar arteries* are derived from the vertebrals near the point of their union, or from the basilar artery; they wind around the upper part of the medulla oblongata, and are distributed to the under surface of the cerebellum, their branches penetrating the fissure between its hemispheres.

The BASILAR ARTERY is the vessel resulting from the union of the two vertebrals; it reaches from the posterior to the anterior border of the pons Varolii, and terminates by dividing into two branches for the cerebrum; it gives off numerous small transverse branches, the most anterior of which are called the *superior cerebellar arteries;* these wind around the crus cerebri on each side, and are distributed to the upper surface of the cerebellum, anastomosing with the inferior cerebellar arteries.

The *posterior cerebral arteries* are the terminal branches of the basilar; they are of large size, and are directed backward around the crura cerebri to the posterior lobes of the cerebrum, which they supply. The arteries are joined near their commencement by the *posterior communicating arteries* from the internal carotid.

The internal carotid artery (p. 52) supplies the anterior and middle cerebral lobes, and at the base of the brain lies just behind the optic nerves.

The *anterior cerebral arteries* are given off from the internal carotids; they penetrate the fissure between the cerebral hemispheres and supply the inner surfaces of the anterior lobes; as the two arteries are about to enter this fissure, they are united by a short transverse branch called the *anterior communicating artery*.

The *middle cerebral arteries*, larger than the preceding, pass outward from the internal carotid to enter the fissure of Sylvius and supply the anterior and middle cerebral lobes. Close to their origin, these arteries give off the *posterior communicating branch* which unites them with the posterior cerebral arteries of the basilar, thus completing the anastomoses of the cerebral arteries, to which the term *circle of Willis* has been applied.

The arachnoid and pia mater, together with the vessels, may now be removed by carefully tearing them away with the forceps; it must be done in such a manner as to respect the origins of the nerves, many of which are small, and so easily confounded with the membranes, that they may be torn away unawares.

The removal of the membranes gives an opportunity to define better than has yet been done, the different parts of the encephalon, their limits, and the subdivisions into which they are separated.

The ENCEPHALON consists of the medulla oblongata, pons Varolii, cerebellum, and cerebrum. The medulla oblongata is the upper part of the spinal cord. The pons Varolii is a broad band of transverse fibres stretching across the upper part of the medulla oblongata. The cerebellum, or little brain, lies beneath the posterior part of the cerebrum, and is composed of two lateral halves or hemispheres. The cerebrum, or large brain, constituting the principal part of the encephalon, is composed likewise of two hemispheres, but is also divided into anterior, middle, and posterior lobes; the anterior and middle lobes are separated by a sulcus called the *fissure of Sylvius;* the limit between the middle and posterior lobes is indicated by a line corresponding to the anterior border of the cerebellum. The convexity of the cerebrum shows no marks of the subdivision into lobes.

ORIGINS OF THE CRANIAL NERVES.

The CRANIAL NERVES, with the exception of the spinal accessory, have their origin in the encephalon and pass through apertures in the skull. The origin of a nerve is described as either *apparent* or *deep*, the former term referring to the point at which it appears superficially, and the latter to its commencement in the nervous substance. The apparent origin of the nerves will alone be given.

The FIRST, or OLFACTORY NERVE, lies in a sulcus on the under surface of the anterior lobe of the cerebrum; it is prismatic in form, and terminates anteriorly in a bulb; posteriorly it is connected with the cerebrum by three roots: an *inner root*, not always apparent, from the inner part of the anterior lobe, a *middle root* from the posterior part of the sulcus in which the nerve lies, and an *external root* from the posterior lip of the fissure of Sylvius.

To see the middle root, the nerve must be reflected and examined from its deep surface.

The SECOND, or OPTIC NERVE, is the largest of the cranial nerves. Anteriorly the nerves of the two sides unite and form the *optic commissure*, or *chiasma;* behind the commissure the nerve is called the *optic tract;* it winds around the crus cerebri, and splits into two parts, which may be traced back to the optic thalamus and the corpora quadrigemina.

The THIRD, or MOTOR OCULI NERVE, arises just in front of the pons Varolii, near the locus perforatus, from the inner side of the crus cerebri.

The FOURTH, or TROCHLEARIS NERVE, the smallest of the cranial nerves, emerges between the cerebrum and cerebellum at the outer side of the crus cerebri, round which it winds. It may be traced to the valve of Vieussens.

The FIFTH, or TRIFACIAL NERVE, springs from the side of the pons Varolii near its anterior border, where it emerges as two roots, separated from each other by a slight interval.

The SIXTH, or ABDUCENS NERVE, springs from the upper part of the corpus pyramidale of the medulla oblongata, close behind the posterior border of the pons Varolii.

The SEVENTH NERVE consists of two trunks, the FACIAL and the AUDITORY. The facial is the smaller of the two and most internal. It arises from the medulla oblongata, between the olivary and restiform bodies, close to the pons Varolii. The auditory trunk arises by distinct filaments from the floor of the fourth ventricle. A small bundle of nervous filaments lies between these two nerves, and is called the *intermediate portion of Wrisberg;* it unites with the facial nerve.

The EIGHTH NERVE consists of three trunks, the GLOSSO-PHARYNGEAL, the PNEUMOGASTRIC, and the SPINAL ACCESSORY; they are all situated along the side of the medulla oblongata.

The *glosso-pharyngeal nerve* is the smallest of the three, and arises by three or more filaments, springing from the restiform body, close to the facial nerve.

The *pneumogastric,* or *par vagum nerve,* arises below the preceding, also from the restiform body, by a series of filaments which are finally gathered into a flat band.

The *spinal accessory nerve* consists of two parts, one being accessory to the pneumogastric and the other spinal. The *accessory* part arises from the spinal cord by a series of minute filaments in a line with the pneumogastric, extending as low as the first cervical nerve; this portion does not unite with the pneumogastric nerve until it gets outside the skull. The *spinal* part is a round cord arising by filaments from the spinal cord, as low down as the sixth cervical nerve. The glosso-pharyngeal, pneumogastric, and spinal accessory nerves converge and meet just below the crus cerebelli.

The NINTH, or HYPOGLOSSAL NERVE, arises by a series of filaments from the sulcus, between the pyramidal and olivary bodies of the medulla oblongata, in a line with the anterior roots of the spinal nerves.

MEDULLA OBLONGATA AND PONS VAROLII.

The MEDULLA OBLONGATA is the upper enlarged portion of the spinal cord. Its shape is pyramidal and its length is about an inch and a quarter; its base is limited by the transverse fibres of the pons Varolii, but its apex blends with the cord and is not as definitely marked. The anterior surface is convex, the posterior is somewhat concave and forms the floor of the fourth ventricle. It is divided into halves by a median longitudinal fissure in front and behind, the anterior ceasing at the pons Varolii in a dilated part called the *foramen cæcum;* the posterior is prolonged into the floor of the fourth ventricle. Each half of the medulla is made up from the three lateral divisions of the spinal cord, indications of which still exist, though they are somewhat differently arranged and changed in form. That portion nearest the longitudinal fissure in front, is called the *anterior pyramid* or *corpus pyramidale;* at about an inch below the pons Varolii the pyramids of the two sides communicate across the anterior longitudinal fissure by what is called a *decussation* of their fibres. The anterior pyramids are the continuation upward of the anterior columns of the spinal cord. Next the anterior pyramids comes the *corpus olivare.* This is an oval projection, separated by a groove on each side from the anterior pyramids in front, and the restiform body behind; its upper end is the most prominent, and does not quite reach the pons Varolii; its lower end is characterized by some transverse fibres arching over its surface, which are called *arciform fibres.* That part of the medulla below the olivary body is continuous with the lateral column of the spinal cord, and is here sometimes called the *lateral tract;* its fibres are continued upward, much diminished in size, between the olivary and restiform bodies. On each side of the posterior longitudinal fissure is the *corpus restiforme.* This is the largest prominence of the medulla, and by its lateral projection gives width to its upper part. The *posterior pyramids* are placed on each side of the posterior longitudinal fissure, lower down than the restiform bodies, and are the continuation upward of the posterior columns of the spinal cord; at the lower part of the medulla they diverge and become blended with the restiform bodies.

The PONS VAROLII, or, as it is sometimes called, the TUBER ANNULARE, is situated above the medulla oblongata and between the hemispheres of the cerebellum; it is of a square shape, the anterior border arching over the crura cerebri, and the posterior over the medulla oblongata. The superficial fibres of the pons Varolii are transverse, and collect together on each side to form the *crura cerebelli;* the deep fibres are longitudinal, and are prolonged upward from the medulla. The *crura cerebri* are large stalk-like bodies emerging from the anterior border of the pons Varolii and dipping into each lateral hemisphere of the cerebrum; as they enter the hemispheres they are crossed transversely by the optic tract.

BASE OF THE CEREBRUM.

Between the crura cerebri on each side and the optic commissure in front is the *inter-peduncular space*, containing the locus perforatus, the corpora albicantia, and the tuber cinereum.

The *locus perforatus*, sometimes called the *pons Tarini*, or *posterior perforated space*, is a layer of whitish-gray substance, between the corpora albicantia in front, and the pons Varolii behind. It gets its name from being perforated by numerous arterial twigs, intended for the interior of the brain.

The *corpora albicantia* are two small round bodies, placed just behind the optic commissure; they are the termination of the crura of the fornix.

The *tuber cinereum* is a mass of gray matter between the corpora albicantia and the optic commissure; it forms the floor of the third ventricle; projecting from it is a small conical body called the *infundibulum*, which is connected with the pituitary body.

The *pituitary body* lies in the sella turcica, and has been examined in connection with the base of the skull (p. 29).

If the two hemispheres of the brain be gently parted anteriorly, the *corpus callosum* will be seen; this is the broad transverse band which unites the two hemispheres of the brain; it has a rounded border in front, called its *genu*, from which it extends backward under the name of the *rostrum*, forming a thin, concave margin, to which is joined a layer of gray substance, called *lamina cinerea*, which unites it posteriorly with the tuber cinereum. Laterally the corpus callosum will be seen extending into the anterior lobes of the lateral hemispheres. A band of white substance, diverging on either side, crosses the substantia perforata to lose itself at the entrance of the fissure of Sylvius. The divergent processes are known as the *peduncles of the corpus callosum*. The corpus callosum will be further examined in connection with the interior of the brain.

The *fissure of Sylvius* is the separation between the anterior and middle lobes; externally the fissure divides into two parts, one of which passes before and the other behind a cluster of convolutions called the *island of Reil*. At the inner extremity of the fissure is a triangular spot, called the *substantia perforata*, or *anterior perforated space*, from being pierced by a number of openings for small arteries; its inner side is continuous with the lamina cinerea.

UPPER SURFACE AND INTERIOR OF THE CEREBRUM.

The brain is to be turned over to bring its convexity uppermost; a roll of cloth should support the whole, and elevate the anterior lobes so as to bring them on a level with the posterior.

The two hemispheres are separated by a median longitudinal fissure, at the bottom of which will be seen the broad white band of the corpus callosum, interrupting the fissure at its central part, but leaving the separation of the hemispheres complete both in front and behind it. The superficies of the hemispheres is marked by tortuous eminences called *gyri*, or *convolutions*, and by intervening depressions called *sulci*. Some of these convolutions have been named, as, for instance, that at the bottom of the median fissure, where the hemisphere rests upon the corpus callosum, which is called the *convolution of the corpus callosum* or *gyrus fornicatus*. The hemispheres of the cerebrum sometimes present a want of symmetry as to size; Bichat believed this to be a condition inconsistent with a perfect performance of the functions of the brain: no more striking proof of the incorrectness of such a view can be found than the fact that the brain of Bichat himself presented this very peculiarity.

The upper part of each hemisphere is to be removed by a horizontal section carried as low down as the convolution of the corpus callosum.

The section thus made will show that each convolution consists, superficially, of a layer of grayish-colored cerebral substance, lying upon a whiter structure constituting the central portion of the section; the gray matter is called *cortical*, and the white *medullary*. If the convexity of a single hemisphere has been cut away, the medullary matter of that side will be seen to have an oval shape; this is called the *centrum ovale minus;* if both have been cut away, a larger medullary surface is exposed, and the two ovals will be found to constitute one large oval, to which the name *centrum ovale majus* is given; the surface of this section, in a recent brain, is dotted with minute bloody points resulting from the division of small vessels.

The CORPUS CALLOSUM is the great commissure of the brain. It is about three inches in length; its fibres are transverse and marked along the central line by a *raphé;* if the convolution called gyrus fornicatus be cut through transversely, and turned one part forward and the other backward, a white, longitudinal band of fibres, lying upon the corpus callosum, will be exposed, called the *striæ longitudinales laterales,* or *covered band of Reil.* In front, the corpus callosum bends round to the base of the brain, the bend being called its *genu;* posteriorly it forms a thick rounded border; on each side it forms the roof of the cavities in the hemispheres called the lateral ventricles, and, by its under surface, is connected with the septum lucidum anteriorly and with the fornix posteriorly.

The *ventricles* of the brain are cavities which exist in its interior; they are five in number. One is found in the central part of each hemisphere; these are called lateral ventricles, and constitute the first and second; the third occupies the middle line of the brain near its under surface; the fourth is between the cerebellum and the posterior surface of the medulla oblongata and pons Varolii; the fifth is in the partition which separates the two lateral ventricles from each other.

The lateral ventricle of either side is displayed by shaving off as much of the corpus callosum and central medullary substance of each hemisphere as constitutes its roof; the escape of a fluid which is ordinarily present, warns the dissector of his arrival in the cavity of the ventricle.

The LATERAL VENTRICLE is a narrow interval extending into the anterior, middle, and posterior lobes of the hemisphere by three prolongations called cornua; the *anterior cornu* is directed outward; the *posterior,* which is small, turns inward, and the *descending, middle,* or *inferior cornu* penetrates in a spiral manner into the middle lobe. This ventricle presents from before backward the following parts for examination, viz:—

Corpus striatum,   Hippocampus major,
Tænia semicircularis, Hippocampus minor,
Optic thalamus,   Septum lucidum,
Choroid plexus,   Fornix.

The *corpus striatum,* a large oval body in the anterior part of the lateral ventricle, is so called from the striated lines of white and gray matter which are seen upon cutting into its substance. The broad end, directed forward, is separated by the septum lucidum from its fellow of the opposite side; its narrow end projects backward outside the optic thalamus.

The *tænia semicircularis* is a narrow band of longitudinal fibres, lying between the corpus striatum and the optic thalamus; in front it joins the pillar of the fornix, and posteriorly is lost in the descending cornu of the ventricle.

The *optic thalamus* is an oblong body, a part of the upper surface of which is alone seen, the rest being covered by the fornix; its posterior portion projects into the descending cornu of the ventricle; its other relations will be subsequently examined.

The *choroid plexus* is a vascular fringe, formed from the pia mater, which extends obliquely across the ventricle; posteriorly it penetrates into the descending cornu, and anteriorly it communicates with the plexus of the other side through the space behind the crura of the fornix as they arch downward, called the *foramen of Monro*, and this may be demonstrated by pulling the plexus gently with a pair of forceps, while the septum lucidum is held aside by the handle of a scalpel.

The *hippocampus major* is the projection which lies at the commencement of the descending cornu, behind the choroid plexus; it follows the direction of the cornu, and its extremity has been likened to the foot of an animal, the two or three indentations, with corresponding elevations, which are found upon it having been named *pes hippocampi*. In order to see this, the cornu must be laid open by a careful division of the external wall of the hemisphere, following a direction backward, outward, and downward, and then forward and inward. The primary letters of the words used to express the curve taken by the cornu have given rise to a mnemonic symbol, BODFI, by which the student may with ease remember its course.

The *hippocampus minor* is the projection which occupies nearly all the space of the posterior cornu. Between the two hippocampi is an elevation called the *eminentia collateralis*.

The corpus callosum is to be cut through transversely at its middle; the anterior half is to be raised by separating it from the fornix and reflected forward; this exposes the septum lucidum and its ventricle. The posterior half is to be raised and reflected backward; the fornix will thus be exposed.

The *septum lucidum* is a thin, triangular, and almost transparent partition between the lateral ventricles; it consists of two layers, inclosing a space which is the *fifth ventricle*. The septum lucidum is so delicate a structure, that if the decomposition of the brain has at all advanced, it will only be seen as it tears away during the gentle lifting of the anterior portion of the corpus callosum, to the under surface of which it is attached superiorly; its lower border corresponds to the rostrum, and its apex to the genu of the corpus callosum; its base is applied to the fornix as it arches downward in front.

The *fornix* is a thin, horizontal layer of white substance, triangular in outline, attached along the median line to the under surface of the corpus callosum, behind the septum lucidum; in front it arches over the foramen of Monro, to form on either side *crura*, which terminate in the corpora albicantia and optic thalami; posteriorly it is continuous with the corpus callosum, and it sends off on each side an expansion called its posterior pillars, which lie along the hippocampus major. It is this part of the fornix which is seen in the floor of the lateral ventricle when first laid open, and the thin edge of which, concealed by the choroid plexus, is called the *corpus fimbriatum* or *tænia hippocampi*. If the fornix be divided anteriorly and reflected backward, there will be seen upon its under surface, between its diverging pillars behind and its crura in front, a triangular space marked by transverse lines, and called the *lyra*. The reflection of the fornix exposes the optic thalami, and it will be seen that it rests upon and covers in these bodies in nearly their whole extent. If we follow the fornix by its free edge or tænia hippocampi into the descending cornu of the ventricle, we shall find, on lifting it from the optic thalamus, that there is a fissure extending downward to the extremity of the posterior cornu, and forward to the foramen of Monro, through which an expansion of the pia mater enters the ventricle, and forms the fringed margin which lies along the tænia hippocampi, called the choroid plexus. This is called the *transverse fissure of the cerebrum*, or *the fissure of Bichat*. The ventricle is made a closed cavity by the union of its lining membrane with that portion of the pia mater which passes through this fissure.

The central part of the pia mater which thus enters the ventricles, lying beneath the fornix, and corresponding to it in shape and extent, is called the *velum interpositum;* its upper surface is in contact with the fornix, and its margin laterally is the choroid plexus; its under surface is the roof of the third ventricle, and it covers the pineal body and a part of each optic thalamus. The velum is traversed along its median line by two veins called *venæ Galeni*, which, commencing at the foramen of Monro, terminate posteriorly at the straight sinus.

The velum interpositum is now to be raised and thrown backward. It must be done with care not to injure the parts beneath.

On the under surface of the velum are the *choroid plexuses of the third ventricle;* continuous in front by the foramen of Monro with the plexus choroides of the lateral ventricle, which they resemble in their structure, they are connected posteriorly with the pineal gland, and this connection is to be remembered in removing the velum, or that body may be torn away with it.

The THIRD VENTRICLE is an interval between the optic thalami; its roof is the velum interpositum, and its floor the tuber cinereum at the base of the brain; in its anterior portion are the descending crura of the fornix and the anterior commissure; in its posterior portion the pineal gland and posterior commissure, covering in the corpora quadrigemina. On its two sides we have the optic thalami, and, crossing from one to the other of these, the middle or soft commissure.

The *anterior commissure* is a round bundle of white fibres, situated just in front of the crura of the fornix, which must be separated in order to display it; it crosses between the corpora striata.

The *middle* or *soft commissure*, frequently torn across in examining the brain, consists of gray matter, and connects the adjacent sides of the optic thalami.

The *posterior commissure* is a flattened white band, connecting the optic thalami posteriorly.

The space between the anterior and middle commissures is called the *foramen commune anterius*, or *foramen of Monro*, and is the medium of communication between the lateral and third ventricles, and of the transmission of the choroid plexuses and their termination, the venæ Galeni. This foramen is also called the *iter ad infundibulum*, from leading down to the funnel-shaped cavity of the infundibulum. (p. 29.)

The space between the middle and posterior commissures

is called the *foramen commune posterius;* it is the point from which a canal, called the *aqueduct of Sylvius*, or *iter a tertio ad quartum ventriculum*, leads backward through the base of the corpora quadrigemina to the fourth ventricle, completing the intercommunication existing between all the ventricles. The two lateral ventricles communicate with each other transversely, and with the third perpendicularly, through the foramen of Munro, and we have just seen how the third and fourth ventricles are connected. It is a disputed point whether the fifth ventricle communicates with the third; if it does, it is by a very narrow orifice at the angle formed by the inferior and posterior borders of the septum lucidum. The weight of authority is against the existence of any aperture at that point.

The *optic thalamus* is a square-shaped body forming part of both the lateral and the third ventricles; it forms a large part of the descending cornu of the former, where it presents two rounded tubercles called *corpus geniculatum externum* and *internum*, and is in relation externally with the corpus striatum and substance of the hemispheres. Anteriorly the thalami are connected with the crura of the fornix as they descend to the corpora albicantia, and posteriorly they are joined by bands from the pineal body and the corpora quadrigemina.

The *pineal body* is a small conical mass of a reddish color and vascular character. situated above, and connected by its base with, the posterior commissure; it lies between the anterior pairs of the corpora quadrigemina; it is also connected with the optic thalami by white bands called its peduncles. The pineal body is hollowed out into a cavity containing, in addition to a viscid fluid, a quantity of gritty matter called *acervulus*, easily detected by rubbing a portion of the mass between the fingers. The pineal body was supposed by Descartes to be the seat of the soul.

The *corpora quadrigemina* are four small, rounded protuberances arranged in pairs and separated by grooves; each pair is situated on the crus cerebri of the same side. The anterior pair, called the *nates*, are the largest; they are oblong in shape and send forward a white band to join the optic thalami. The posterior pair, called the *testes*, are rounder and whiter than the preceding, and are also connected with the thalami by a white band. The *iter a tertio ad quartum ventriculum* passes longitudinally through the base of the corpora quadrigemina.

By drawing back the cerebellum two large white cords will be seen extending from the corpora quadrigemina to the cerebellum; these are termed the *processus e cerebello ad testes*, or *superior peduncles* of the cerebellum. Between them is stretched a thin layer of cerebral substance, covering in the passage from the third to the fourth ventricle, as well as a part of the fourth ventricle itself; this is the *valve of Vieussens*.

CEREBELLUM.

The cerebellum is to be detached from the remains of the cerebrum by carrying the knife through the optic thalami, so that the cerebellum, with the corpora quadrigemina, pons Varolii, crura cerebri, and medulla oblongata, shall remain united together.

The CEREBELLUM, or little brain, lies beneath the posterior lobes of the cerebrum, from which, in the cranial cavity, it is separated by the tentorium; its diameter is greatest from side to side, and it is incompletely divided into two lateral hemispheres. Its surface is marked by plates or laminæ instead of convolutions, and between these are fissures lined by pia mater; the deeper of these fissures break the hemispheres into imperfectly marked lobes.

The two hemispheres are united on the upper or cerebral aspect by a central, constricted isthmus, called the *superior vermiform process;* in front of this is a notch which encircles the corpora quadrigemina posteriorly, called *incisura cerebelli anterior;* behind the isthmus is another notch called *incisura cerebelli posterior*. Upon the under surface of the cerebellum the depression which receives the medulla oblongata is called the *vallecula*, and at the bottom of this is an isthmus, corresponding to that connecting the hemispheres on the superior surface, and called the *inferior vermiform process*.

The cerebellum is united to the rest of the encephalon by a large stalk-like process on each side. This process subdivides into three rounded cords, called peduncles. The *superior peduncle*, or *processus e cerebello ad testes*, is directed forward to the corpora quadrigemina, and forms the anterior part of the lateral boundary of the fourth ventricle; between these peduncles is the valve of Vieussens. The *middle peduncle*, or *processus ad pontem*, is commonly called the *crus cerebelli*, and is the largest of the three peduncles. Its fibres begin in the lateral part of the cerebellum, and are directed forward to the pons Varolii, of which they form the transverse fibres, uniting with the fibres of the crus of the opposite side. The *inferior peduncle*, or *processus ad medullam*, passes downward to the medulla oblongata and forms part of the restiform body; it is also the inferior portion of the lateral boundary of the fourth ventricle.

The FOURTH VENTRICLE is a space between the cerebellum and the posterior aspect of the medulla oblongata and pons Varolii; it has the shape of a lozenge, the points of which are directed upward and downward, the upper extending as high as the superior border of the pons Varolii, and the lower to a level with the inferior part of the olivary body. The lower half of this space has been called the *calamus scriptorius*, from a fancied resemblance to the tip of a writing pen; its lateral boundaries are formed by the superior peduncles of the cerebellum above, and by the restiform bodies below; its roof is formed by the valve of Vieussens and the sides of the peduncles with which it is connected, and by the under part of the superior vermiform process; its floor corresponds to the posterior surfaces of the medulla and pons Varolii. Along the centre of its floor is a median groove, ending at the point of the calamus, and on each side of the groove a fusiform elevation called the *eminentia teres*. The eminentia teres is crossed by white streaks, varying in their arrangement and not always recognizable, called *lineæ transversæ*. The fourth ventricle communicates with the third ventricle through the *iter a tertio ad quartum ventriculum*, and by this canal the lining membrane of the other ventricles is prolonged into the fourth.

The arachnoid and pia mater stretch across between the medulla oblongata and cerebellum, at the lower part of the ventricle, under the name of the *valve of the arachnoid*. On the inner surface of these membranes is a vascular fold, called the *choroid plexus of the fourth ventricle*. The ventricles communicate with the sub-arachnoid space of the encephalon and spinal cord, through an aperture in this membrane just above the calamus scriptorius.

A vertical section of the cerebellum shows that the laminæ of its hemispheres are composed of an external layer of gray matter and an internal layer of white medullary substance, the white part being derived from a large medullary centre in the interior of each hemisphere, which in one direction gives rise to the peduncles of the cerebellum, and in the other, diverging like the branches of a tree, enters by small offsets into each lamina; the appearance thus presented is called the *arbor vitæ cerebelli*.

A portion of the medullary centre of the cerebellum will be found inclosed by a waved or dentated gray line; this is called the *corpus rhomboideum*, and a similar appearance, bearing the same name, may be found in the olivary body of the medulla, on slicing it obliquely in its long diameter.

By dividing transversely one of the crura cerebri, a large mass of gray matter, called the *locus niger*, will be found in the centre of the section.

DISSECTION XIV.

THE INTERNAL AUDITORY APPARATUS.

In order to dissect the ear, a recent temporal bone should be obtained; the soft parts, already examined, should be removed. The squamous portion of the bone is to be got rid of by a vertical cut through the root of the zygoma, and the anterior wall of the osseous portion of the auditory canal is to be cut away with bone forceps; this will expose the tympanic membrane. It is extremely difficult, however, to get an idea of the parts contained within the temporal bone; it is only upon a model, or one of the French preparations, that it can be done effectually. The dissector may find his work facilitated by softening a recent bone in acid, but with every aid the examination will require a very considerable amount of skill and knowledge.

The internal portions of the auditory apparatus are all contained within the temporal bone, which also forms a part of the external auditory canal. The tympanic membrane is placed at the bottom of this canal, and separates the external portion of the ear from that constituting the sentient structure.

The *osseous portion of the auditory canal* is about three-quarters of an inch in length; it is furnished with a rough lip, called the *processus auditorius*, into which the fibro-cartilage of the external ear is inserted; its lower wall describes a curve, the convexity of which is directed upward, and the summit of which hides the lower part of the tympanic membrane from view, in examinations with the speculum auris.

The *membrana tympani* is a thin, oval, and semi-transparent membrane, attached by its circumference to a groove in the bone at the inner end of the auditory canal; its surface is convex internally, and concave externally; it is placed at an angle of 45° with the floor of the canal. One of the little bones of the ear, the malleus, is attached to the upper part of its internal surface, which is also crossed by the chorda tympani nerve. In the fœtus this membrane is connected with a separate, osseous ring, the *tympanic bone*, which subsequently unites with the rest of the temporal bone.

MIDDLE EAR.

The tympanum is behind the tympanic membrane, and is exposed by removing the bone which forms the roof of the cavity, in such a way as to preserve as perfectly as possible the membrane and the chain of little bones within it.

The TYMPANUM, or MIDDLE EAR, has the form of a round, flat box, placed on edge; the following points are to be noticed within its cavity. Upon the surface, opposite the membrana tympani, a central projection, called the *promontory;* above and below this, two apertures opening into the labyrinth; the superior one is called the *fenestra ovalis*, and the lower the *fenestra rotunda;* the fenestra rotunda is closed by a thin membrane, called the *secondary membrana tympani*. The anterior boundary of the tympanic cavity is formed by the membrana tympani and a part of the surrounding bone; above the membrane may be seen the *fissura Glaseri*, occupied by the long process of one of the small bones of the ear, the malleus, and by a small muscle, the laxator tympani; crossing the membrana tympani at its upper part is the *chorda tympani nerve*. This nerve is a long but slender branch of the facial; it arises about a quarter of an inch from the stylo-mastoid foramen, and passes forward to the tympanum, entering that cavity just below the pyramid; it then crosses the handle of the malleus and the membrana tympani to a small foramen on the inner side of the Glaserian fissure, through which it passes, and, emerging, unites with the gustatory nerve.

It is to be noticed that the circumference of the tympanic cavity presents a rough surface, separated superiorly from the cranial cavity, and inferiorly from the jugular fossa, merely by a thin, osseous plate; this is an anatomical fact of importance in connection with fractures of the petrous portion of the temporal bone, which, if at this point, would lead to serious complications. At the posterior part of the circumference is an orifice leading to the mastoid cells; below this aperture is a projection called the *pyramid*, from which a small spiculum of bone extends to the promontory; the apex of this pyramid is open, and from it emerges the stapedius muscle; arching up from the pyramid, above the fenestra ovalis, is a ridge of bone marking the aqueduct of Fallopius. At the anterior part of the circumference are the apertures of two canals, the upper one of which contains the tensor tympani muscle, and the lower one the Eustachian tube. Between these two canals is a thin, osseous lamina, called the *processus cochleariformis*.

The *Eustachian tube* is the channel through which the tympanic cavity communicates with the fauces; it is an inch and a half in length, and is partly osseous and partly cartilaginous in structure; its course in the temporal bone is along the angle of union of the squamous and petrous portions, external to the aperture which contains the carotid artery. Its cartilaginous portion has been already described (p. 62).

The tympanum contains three small bones. They extend in a line across the cavity, and are named malleus, incus, and stapes.

The *malleus*, so called from its supposed resemblance to a hammer, is large at one end (head), and small and tapering at the other (handle); it has two processes, a long and a short; the short process springs from the root of the handle, and the long from a point just above it; the long process extends into the fissura Glaseri; the handle is attached to the membrana tympani; upon the side of the head of the malleus is an articulating surface which unites it with the next bone, the incus.

The *incus*, or anvil-shaped bone, consists of a body and two processes; the body is concave, to receive the head of the malleus; the two processes, long and short, shoot out from the side of the body, opposite its articulating surface. The long process terminates in a rounded extremity called the *orbicular* process, sometimes described as a separate bone, under the name of *os orbiculare*. The orbicular process unites the incus with the third bone, the stapes.

The *stapes*, resembling in its shape a stirrup, has a base and two sides which unite to form a head. The head is marked by a superficial depression which articulates with the orbicular process of the incus; the base, which is a thin, osseous plate, is fixed over the fenestra ovalis.

These bones are maintained in their position by minute ligaments, which unite them to each other and to the surrounding walls. Three muscles, also, are in connection with the chain of bones.

The *tensor tympani*, the largest of the three muscles, is contained in an osseous canal, the orifice of which has already been spoken of as lying above that of the Eustachian tube. It arises from the walls of this canal, and terminating in a tendon which turns round the processus cochleariformis as a pulley, is inserted into the inner border of the handle of the malleus, at its base.

The *stapedius muscle* arises from a canal, in the interior of the pyramid; it ends in a small tendon which is inserted into the neck of the stapes.

The *laxator tympani* arises from the spinous process of the sphenoid bone, and is continued by a long tendon through the fissura Glaseri, to be inserted into the neck of the malleus, above its long process.

INTERNAL EAR.

The INTERNAL EAR, or LABYRINTH, is composed of an osseous and a membranous portion. The osseous labyrinth consists of the vestibule, the semicircular canals, and the cochlea; the membranous labyrinth, of two sacs, called the utricle and the saccule.

The vestibule is exposed by enlarging the fenestra ovalis.

The VESTIBULE is an oval-shaped cavity, situated behind the posterior wall of the tympanum; within it are numerous openings; five of these belong to the three semicircular canals, and a larger one, in front of the others, leads to the cochlea. A vertical ridge, called the *crista*, traverses the inner wall; below this is a small circular depression, the *fovea hemispherica*, perforated by minute orifices which transmit nervous filaments; this depression corresponds to the bottom of the meatus auditorius internus. On the roof of the vestibule is another oval depression called the *fovea semi-elliptica*. Behind the crista, and near the common opening of two of the semicircular canals, is the internal opening of the *aquæductus vestibuli*, the other orifice of which is on the posterior surface of the petrous portion of the temporal bone; through it pass a small artery and vein.

The semicircular canals, above and behind the vestibule, may be followed out from their orifices in the vestibule by a file and strong knife; some of the instruments used by dentists are applicable to this and other parts of the dissection of the internal ear.

The SEMICIRCULAR CANALS, three in number, are of unequal length, though each forms more than half a circle; they communicate by both ends with the vestibule; as two of them, however, blend together, they have but five openings into that cavity. Each canal has a dilated extremity, called its *ampulla*. From their different directions the canals have been named superior, oblique, and horizontal. The *superior* canal crosses the upper part of the petrous bone transversely, on the anterior surface of which, within the cranium, it makes a marked projection; its ampulla is at the outer end of the canal; its inner end joins with the oblique canal. The *oblique* canal is directed backward toward the posterior surface of the temporal bone; its inner end is in common with that of the superior canal, and its outer end is furnished with an ampulla. The *horizontal* canal is the shortest of the three, and has separate orifices; it lies in the substance of the bone, nearly on a level with the fenestra ovalis; its ampulla is on the outer side, close above that aperture.

The cochlea is anterior to the vestibule. To expose it, the surface of bone forming the promontory of the tympanum must be filed or cut away; the surface of the petrous bone above it should also be removed.

The COCHLEA is conical in form, with its base turned toward the meatus auditorius internus. It resembles a snail-shell in construction, consisting of a tube wound spirally round a central part or axis; the tube makes two turns and a half round the axis, and terminates in a closed extremity, called the *cupola;* the axis, or *modiolus*, is the bony centre included within the coils of the spiral tube; its shape is conical, and its size diminishes as it reaches the apex of the cochlea. Winding round the axis is a thin osseous plate, called the *lamina spiralis*, which is projected into the spiral tube, and forms part of a septum, completed by a membrane, which divides the tube into two passages; these, however, communicate at the apex of the cochlea by a foramen, called the *helicotrema*. The lamina spiralis terminates at the apex of the cochlea, in a hook-shaped process, called the *hamulus*. The two divisions of the spiral tube are called *scalæ*, that nearest to the apex being the *scala vestibuli*, and the other the *scala tympani;* the scala vestibuli opens by an oval aperture into the anterior part of the vestibule; the scala tympani is shut off from the vestibule, and terminates at the fenestra rotunda, which is closed by a membrane. Communicating with the scala tympani is the *aquæductus cochleæ*, which opens on the under surface of the petrous portion of the temporal bone, and transmits a vein from the cochlea to the internal jugular vein.

The vestibule, the semicircular canals, and the scalæ of the cochlea are lined with a delicate fibro-serous membrane, which secretes a thin fluid called the *liquor Cotunnii*.

The MEMBRANOUS LABYRINTH is composed of sacs, over which the auditory nerve is expanded, and which contain a fluid called the *liquor Scarpæ;* these sacs are two in number, the utricle and the saccule, and are a perfect counterpart, in form, of the vestibule and semicircular canals. The *utricle* is situated in the posterior and upper part of the vestibule, opposite the fovea semi-elliptica in the roof. The *saccule* is smaller than the utricle, and is placed in front of the latter, in the hollow of the fovea hemispherica. The membranous tubes for the semicircular canals are prolongations from the utricle. In these sacs will be found two small calcareous concretions, called *otoconites*. The membranous labyrinth floats in the liquor Cotunnii.

The auditory nerve divides into two branches in the meatus auditorius internus, called the cochlear and the vestibular.

The *cochlear branch* enters by numerous filaments at the base of the modiolus, and these, bending outward, are distributed upon the lamina spiralis.

The *vestibular branch* divides into three filaments, one of which goes to the utricle and to the membranous labyrinth in the ampulla of the superior and horizontal canals; the second to the saccule, and the third to the membrane in the ampulla of the oblique canal.

The internal ear is supplied by the *internal auditory artery*, a minute branch of the basilar artery, which enters the internal meatus with the auditory nerve, and divides into two branches, one for the cochlea and one for the vestibule.

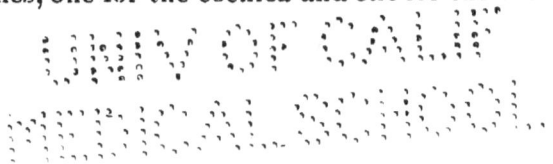

PART SECOND.

ANATOMY OF THE UPPER EXTREMITY, THORAX, AND BACK.

DISSECTION I.

PECTORAL AND DELTOID REGION.

A large block is to be placed under the subject so that the scapulæ shall rest upon it; the arm should extend at an obtuse angle with the body; the dissector stands upon its inner side. An incision is to be made, commencing at the third rib (a longitudinal incision along the median line of the sternum is supposed to have been already made, when the subject was injected), and continued in a straight line to about the middle of the arm. The two flaps of skin thus marked out, with the fat and cellular tissue beneath, are then to be successively reflected upward to the clavicle, and downward to the limits of the origin of the pectoralis major. This dissection brings the following parts into view:—

The PECTORALIS MAJOR MUSCLE arises from the sternal half of the clavicle, half of the sternum in its whole length, the cartilages of all the true ribs, except the last, and from the aponeurosis of the external oblique muscle of the abdomen; from which a muscular slip is occasionally added, which slip may continue separately from the pectoralis, and have a distinct insertion into the humerus. From this origin the fibres converge to form a flat tendon, inserted into the external ridge of the bicipital groove of the humerus, with this peculiarity, viz: that the fibres are rolled upon themselves, in such a way that those of the upper portion of the muscle are inserted into the lower part of the ridge, and those of the lower portion into the upper part. That part of the muscle arising from the clavicle is separated from that arising from the sternum by a cellular interspace, the size of which, as well as the extent of surface occupied by the sternal origin, varies according to the muscular development, or degree of emaciation presented by each individual subject. Near the sternal origin, the pectoralis major is perforated by a number of arterial twigs which come from the internal mammary artery, and, piercing the intercostal spaces, supply the muscle and the integument.

The mammary gland will have been removed with the integument; it is to be examined by removing the fascia and cellular tissue covering its base. The tubuli of the nipple may be demonstrated by the insinuation of a bristle into their orifices.

The MAMMARY GLAND rests upon the pectoralis major muscle, and is separated from it only by a layer of fascia. In the male subject the gland is rudimentary; in the female it varies in size and development. It consists of lobules and lobes, from which issue *lactiferous ducts*, connected together by fibrous and adipose tissue. The nipple is a conical prominence projecting from the centre of the gland, at which open the *tubuli galactophori*, about twenty in number, formed by the union of the converging lactiferous ducts. The nipple is surrounded by an areola or colored ring, the shade of which is influenced by complexion, and upon which numerous papillæ will be noticed.

The pectoralis major and the deltoid muscles are separated from each other by a groove in which lies the cephalic vein, ascending from the elbow, and an artery, the descending branch of the thoracica acromialis.

The deltoid muscle covers the shoulder and cannot be exposed in its totality until the subject be turned upon its face. It is composed of coarse muscular bundles, difficult of dissection from the impossibility of making them tense; this is best effected by bringing the arm down to the side and rotating it firmly inward.

The DELTOID MUSCLE arises from the external third of the clavicle, from the lower border of the acromion process, and from the inferior border of the spine of the scapula, as far as the triangular space which terminates it, and is inserted into the rough triangular eminence on the outer aspect of the shaft of the humerus. The anterior fibres cover the tendon of the pectoralis major and are in contact with the upper part of the biceps; posteriorly they are bound down by fascia. This muscle covers the convex head of the humerus and the insertions of the scapular muscles into the tuberosities of that bone. A bursa, more or less apparent, lies between these tendons and the under surface of the deltoid muscle.

The dissection of these two muscles completed, the pectoralis major may be divided through its middle and the two ends reflected; in so doing, some offsets of the thoracica acromialis and superior thoracic arteries will be cut across, as well as the short thoracic nerve, given off from the brachial plexus just below the clavicle. On clearing away the fat and cellular tissue, the pectoralis minor muscle comes into view.

The PECTORALIS MINOR MUSCLE arises by three distinct tongues from the third, fourth, and fifth ribs, and is inserted into the anterior border of the coracoid process of the scapula, in common with the short head of the biceps and coraco-brachialis muscles.

Connected with the upper border of the pectoralis minor and inserted into the first rib and the coracoid process, and to the clavicle between these points, is a dense fascia called the *costo-coracoid membrane*. When this is removed it will be found to have concealed a small muscle called the subclavius.

The SUBCLAVIUS MUSCLE arises by a short thick tendon from the cartilage of the first rib; its fibres are directed outward and are inserted into the under surface of the clavicle for more than half its length; this muscle receives a muscular nerve from the brachial plexus behind the clavicle.

AXILLA.

The dissection pursued thus far will have exposed a large space filled with fat, cellular tissue and lymphatic glands, lying behind the pectoralis minor, and through which pass the axillary artery, with its vein, and the brachial plexus of nerves; this space is called the AXILLA. The pectoralis minor forms but a small part of its anterior boundary, which is completed by the pectoralis major; the posterior boundary is formed by the flat tendon of the latissimus dorsi, by the teres major and sub-scapularis muscles; internally it is limited by the serratus magnus and externally by the biceps and coraco-brachialis muscles.

Keeping the arm well extended, patience and a little ingenuity in hooking aside the branches of arteries and filaments of nerves, to make room for the scalpel to reach others, will accomplish the display of the axillary artery and its branches. The brachial plexus of nerves, surrounding the artery, will, to a certain extent, be prepared in this dissection. Veins which it is impossible to avoid dividing should be tied, so that the blood oozing from them may neither soil nor obscure the dissection. In a locality like this, the dissector will find the *scissors* a most effective and serviceable instrument.

The AXILLARY ARTERY is that portion of the artery destined to the upper extremity, intervening between the subclavian and brachial arteries, and extending from the outer border of the first rib to the lower margin of the tendon of the latissimus dorsi and teres major; it passes through the axilla nearer to its anterior than its posterior border. The *axillary vein* lies in front and at the inside of the artery, and is formed by the union of the basilic vein of the arm and the venæ comites of the brachial artery; it receives the cephalic vein near the clavicle, and smaller veins, from the muscles, or which accompany the branches of the axillary artery, enter it at various parts of its course.

The BRACHIAL PLEXUS, formed from the anterior branches of the last four cervical, and the first dorsal nerves (p. 59), lies upon the outer side of the axillary artery at its upper part; lower down it surrounds the artery, giving off its branches at various points. Some of these will be described with the dissection of the arm; those which are given to the thoracic and scapular muscles will now be enumerated, though a further reference to them will be made with the dissection of those muscles. These nerves are six in number, viz:—

Superior Muscular,   Supra-scapular,
Short Thoracic,   Sub-scapular,
Long Thoracic,   Inferior Muscular.

The *superior muscular* supply the subclavius, rhomboidei, and levator anguli scapulæ muscles; they are branches of the fifth cervical nerve, and are given off from the plexus behind the clavicle.

The *short thoracic*, two in number, supply the pectoralis major and minor muscles; they are given off from the plexus at a point parallel with the clavicle.

The *long thoracic* passes down behind the axillary vessels and plexus, ramifies on the side of the thorax, and is distributed to the fibres of the serratus magnus muscle exclusively; it is a branch of the fourth and fifth cervical nerves.

The *supra-scapular* passing outward, goes through the supra-scapular notch, and supplies the supra- and infra-spinatus muscles; it is a branch of the fifth cervical nerve.

The *sub-scapular* are two in number, and are both distributed to the sub-scapularis muscles; one of them comes from the plexus above the clavicle, and the other from the posterior aspect of the plexus within the axilla.

The *inferior muscular* consist of one or more branches, distributed to the teres major and latissimus dorsi muscles; they are given off from the lower part of the plexus.

The axillary artery gives off seven branches which vary extremely, however, as to their precise points of origin. In one case in ten, it gives off a larger branch than usual, and this may be either the radial, the ulnar, the interosseous, or a trunk from which arise the sub-scapular, the circumflex, and profunda arteries. The regular axillary branches are—

Thoracica Acromialis, Thoracica Axillaris,
Superior Thoracic, Sub-scapular,
Inferior Thoracic, Anterior and Posterior Circumflex.

The *thoracica acromialis* emerges in the space above the border of the pectoralis minor muscle and divides into branches, which are directed inward to the two pectoral muscles and outward to the deltoid; from the deltoid set a small twig, called the *inferior acromial*, passes down beside the cephalic vein in the interspace between the pectoralis major and deltoid muscles.

The *superior thoracic* (sometimes given off by the preceding artery), passes along the upper border of the pectoralis minor, distributing branches to the pectoralis major and mammary gland, and inosculates with the branches of the internal mammary, emerging between the costal cartilages near the sternum.

The *inferior thoracic* or *external mammary* (sometimes given off from the thoracica acromialis or the sub-scapular), passes along the lower border of the pectoralis minor and is distributed to the pectoralis major, the mammary gland, and the serratus magnus, anastomosing with the superior thoracic and with external branches of the intercostal arteries.

The *thoracica axillaris* is a small artery (arising frequently from one of the other branches) distributed to the nerves and arteries of the axilla.

The *sub-scapular* is the largest branch of the axillary artery; it passes along the lower border of the sub-scapularis muscle, supplying that muscle as well as those of the lower border of the scapula, and gives off a dorsal branch which, passing under its edge, is distributed to the posterior surface of that bone; this artery and its branches inosculate with the posterior and supra-scapular arteries, branches of the subclavian.

To obtain a view of the two remaining branches, the deltoid muscle must be divided and its two ends reflected; in so doing a few small branches of the posterior circumflex artery will be cut across.

The *anterior circumflex* is a small artery, passing beneath the coracobrachialis muscle and the short head of the biceps, to wind around in front of the head of the humerus: it sends a small branch along the bicipital groove to the shoulder-joint.

The *posterior circumflex*, much larger than the anterior, passes backward beneath the head of the humerus, giving branches to the shoulder-joint, and also, after emerging on the other side of the bone, to the deltoid muscle. These two last-named arteries, together with the profunda, are frequently given off from a common trunk, instead of from the axillary artery.

DISSECTION II.

FRONT OF THE ARM.

An incision is to be made down the middle of the arm, to a short distance below the bend of the elbow, care being taken not to divide the cutaneous veins or nerves, which near the elbow are quite superficial. The anterior portion of the arm is to be cleared from cellular tissue, the numerous nerves, arteries, and veins being carefully managed while the muscles are isolated from each other and the surrounding parts. The insertions of the biceps and brachialis anticus muscles cannot be examined until a subsequent stage of the dissection.

In preparing the front of the arm two large veins will have been traced from the elbow upward; these are the cephalic and basilic.

The CEPHALIC VEIN ascends upon the outer side of the biceps muscle to the space between the pectoralis major and deltoid muscles, along which it passes to terminate beneath the clavicle in the axillary vein.

The BASILIC VEIN, superficial at the lower part, pierces the fascia near the middle of the arm, and, accompanying the artery at its inner side, becomes, in the axilla, the axillary vein.[1]

The BICEPS MUSCLE constitutes the bulk of the arm in front. It arises by two heads, a short and a long; the short head, in common with the coraco-brachialis, arises from the coracoid process of the scapula; the long head arises by a round and slender tendon from the upper margin of the glenoid cavity of the scapula; this passes over the head of the humerus through a special synovial sheath and enters the bicipital groove; emerging from this, it expands into a broader tendon from which the muscular fibres take their origin. These two heads uniting form the belly of the muscle, which, terminating in a flattened tendon,

---

[1] The first two letters of the word "biceps," that being the muscle with which this vein is in relation, will give the student a mnemonic aid to retain in his mind the relative position of the basilic and cephalic veins. (B. I. basilic, internal.)

penetrates between the muscles of the forearm to be inserted into the tubercle of the radius. An aponeurotic expansion continuous with the fascia of the forearm, is given off at the elbow from the outer side of the tendon of insertion, protecting the brachial artery which lies just beneath it. The biceps sometimes has a *third head*, which consists of a bundle of muscular fibres from the front of the humerus, usually connected with the brachialis anticus, and which unites with the lower part of the belly of the muscle on its inner side. Two or three muscular branches of the brachial artery enter this muscle. The inner border of the biceps, at its middle, is the guide to the brachial artery, which lies just within its edge and inside the median nerve.

The CORACO-BRACHIALIS MUSCLE lies at the inside of the biceps; it arises from the coracoid process, between the pectoralis minor and the short head of the biceps, and is inserted into a rough line on the inner side of the middle of the humerus.

The BRACHIALIS ANTICUS MUSCLE lies beneath the biceps on the lower half of the arm; it arises from the humerus by muscular fibres which embrace the insertion of the deltoid, and is inserted into the coronoid process of the ulna. Its insertion cannot be fully seen until the forearm is dissected.

The BRACHIAL ARTERY extends from the lower border of the conjoined tendons of the latissimus dorsi and teres major to the bend of the elbow, where it divides into the radial and ulnar arteries; it lies along the inner border of the coraco-brachialis and biceps muscles, and is superficial in nearly its whole extent; the basilic vein lies in front of the artery and the ulnar nerve along its inner side; the median nerve lies first upon its outer side, then crosses it and descends to the elbow on its inner side.

The normal condition of this trunk may be varied by what is called its "high division;" that is, instead of originating at the usual point, the radial, ulnar, or interosseous arteries, one or all, are given off higher up, along the course of the brachial or even of the axillary artery. Sometimes the radial and ulnar, when thus given off, are connected by a transverse branch. Occasionally the brachial artery descends with the median nerve to a point near the inner condyle, where it turns around a prominence of bone[1]

[1] Called the "supra-condyloid process" of the humerus.

which is occasionally present, and regains its usual position. This anomaly is analogous to the ordinary distribution of the vessel in some carnivorous animals, in which it passes through a foramen in the humerus, a short distance above the inner condyle.

The brachial artery has three branches.

        Superior and Inferior Profunda,
        Anastomotica Magna.

The *superior profunda* is given off close below the tendon of the latissimus dorsi; it passes under, and winds around the humerus, to reappear in the muscular interspace on the outer side of the brachialis anticus, where it inosculates with the radial recurrent artery. The profunda artery sends a branch down the posterior aspect of the humerus, to the articulation, and this inosculates with the recurrent branch of the posterior interosseous artery.

The *inferior profunda* is a small branch given off about the middle of the arm; it passes downward, penetrates behind the inner condyle, and inosculates with the posterior ulnar recurrent. This artery often arises from the preceding, or is altogether wanting.

The *anastomotica magna* is given off upon the inside of the arm just above the elbow; it runs transversely inward through the inter-muscular septum to the hollow between the olecranon and the inner condyle, where it inosculates with the inferior profunda and the posterior ulnar recurrent branch. One of its muscular offsets forms an arch across the back of the humerus with a branch of the superior profunda.

If the various branches into which the brachial plexus of nerves divides, and which are distributed to the arm and forearm, have been carefully followed out, as far as the elbow, in connection with the previous dissection, they will be found to be seven in number, viz:—

External Cutaneous,	Circumflex,
Internal Cutaneous,	Musculo-spiral,
Lesser Internal Cutaneous,	Median,
Ulnar.	

The EXTERNAL, or MUSCULO-CUTANEOUS, or PERFORANS CASSERII NERVE, arises in common with the external head of the median nerve; it supplies the coraco-brachialis, which it pierces (hence its name of *perforans*), and the biceps and brachialis anticus, and passes between these two muscles to appear on the outer side of the elbow, where it divides into two branches, both supplying the integument, the external down to the back of the hand, and the internal as far as the wrist anteriorly.

The two following nerves are extremely liable to be removed with the flap of integument, their small size and superficial position leading to their division as they lie unnoticed in the cellular tissue.

The INTERNAL CUTANEOUS NERVE arises in common with the internal head of the median nerve; it passes down the arm by the side of the basilic vein, giving off cutaneous filaments in its course, and divides into two principal branches, both of which are distributed to the integument of the forearm along its inner and anterior aspect.

The LESSER INTERNAL CUTANEOUS NERVE, or NERVE OF WRISBERG, is the smallest nerve of the arm; it is found inside of the internal cutaneous nerve, and is distributed to the integument of the lower and inner part of the arm. Just below the axilla, this nerve communicates with the *intercosto-humeral nerve*, a branch of the second intercostal nerve, stretching across from the thoracic parietes to the axilla. A second intercosto-humeral nerve, from the third intercostal, sometimes exists.

The CIRCUMFLEX NERVE is a large trunk which crosses the tendon of the sub-scapularis to pass directly underneath and around the head of the humerus, in company with the posterior circumflex artery; it gives branches to the deltoid, the neighboring integument, and to all the muscles of the scapular region. Its sudden turn, and the shortness of its trunk before disappearing behind the humerus, sometimes embarrass the dissector in his search for this branch of the brachial plexus.

The MUSCULO-SPIRAL NERVE arises in common with the circumflex; it passes behind the brachial artery, and winds spirally round the humerus, in company with the superior profunda artery, to reach the outer side of the arm, where it lies deep between the brachialis anticus and supinator longus muscles, and divides into two branches—radial and posterior interosseous.

The MEDIAN NERVE arises by two heads, which, in the axilla, embrace the brachial artery; these form a trunk of large size which lies at first on the outer side of the vessel, afterward on its inner side, and descends without any branches to the bend of the elbow, where it gives off some muscular branches and the anterior interosseous nerve; it then continues down the forearm to the hand.

The ULNAR NERVE arises in common with the internal head of the median, and internal cutaneous nerves, and descends without branches upon the inner side of the brachial artery, to the hollow between the inner condyle and olecranon, where it gives off muscular and articular branches, and continues down the ulnar side of the forearm to the hand.

BEND OF THE ELBOW.

The veins at the front of the elbow should be studied, with reference to the operation of *venesection*. A ligature, placed high up around the arm of a living person, affords, in many respects, a better opportunity for studying them than is found upon the dead subject. They are generally irregular, and rarely correspond with the description given in books; certain relationships between them and the artery and nerves, should, however, be carefully noted.

Three superficial veins return the blood from the forearm—viz: RADIAL, ULNAR, and MEDIAN—the situations of which are indicated by their names. The median vein divides at the elbow into two short branches, which unite respectively with the ulnar and radial veins; the internal branch is called the median basilic, the external the median cephalic; the basilic vein being the continuation of the median basilic and ulnar, and the cephalic of the median cephalic and radial veins. The disposition of these veins at the elbow may be compared to the letter M; the middle angle of that letter representing the division of the median vein, the superior lateral angles corresponding to the commencement of the cephalic and basilic veins, formed by the union of the median cephalic and median basilic with the radial and ulnar veins, which are represented by the limb upon each side of the letter.

The MEDIAN BASILIC VEIN crosses the brachial artery, being separated from it only by the aponeurotic slip given to the fascia of the forearm from the tendon of the biceps. Branches of the internal cutaneous nerve pass both in front and behind this vein.

The MEDIAN CEPHALIC VEIN, smaller than the median basilic, passes outward along the fold of the elbow, somewhat less superficially; the branches of the external cutaneous nerve pass beneath the vein.

Notwithstanding that, in bleeding, the operator usually selects the largest vein, it will be seen that the median cephalic presents the more favorable conditions than the median basilic, being away from the artery and above the nerve, while the latter, though ordinarily the largest, is not only almost in contact with the artery, but surrounded by nerves. This dissection will at least show the importance of exploring the arm for the pulsations of the artery with reference to its relation to the vein, whichever it may be that it is proposed to open, if the dangers of traumatic aneurism would be avoided.

In the dissection of the elbow, a lymphatic gland will be found with a considerable degree of constancy, just above the inner condyle; practically important, as being sometimes enlarged and inflamed from wounds or ulcerations of the hands or fingers, and almost constantly so in cases of constitutional syphilis.

DISSECTION III.

STERNAL REGION.

The further dissection of the upper extremity is now to be relinquished until the thorax and its contents have been examined. It should be carefully wrapped in a bandage and kept constantly damp until its dissection is resumed.

The anterior wall of the thorax is to be removed by dividing upon each side the costal cartilages and intercostal muscles, close to the ribs. If the lungs be free from adhesions, and the pleural cavity has not yet been opened, the student will hear, as he first opens into it, the whistle of the air as it enters the vacuum previously existing, and will see the lungs collapse under the atmospheric pressure to which they are then first subjected. If the dissection of the muscles attached to the upper part of the sternum and clavicle is completed, the whole of the sternum may be removed; otherwise a portion must be left by sawing it across, an inch below its summit. The segment included in the incisions made is to be lifted, first by one and afterward by both of its lower angles, and the cellular and muscular tissues divided which attach it to the parts beneath. Its separation at the first rib and from the clavicle is a little difficult, unless properly performed; the knife, as it approaches the first rib, should be directed obliquely outward till its cartilage is divided, then turning at right angles to this incision, it is to be carried inward, gradually describing a curve with its concavity outward, through the sterno-clavicular articulation. Every autopsy which the student attends affords him opportunity of studying nearly all the parts about to be described.

Upon the inside of the *plastron*, as the segment thus removed is called, beneath a layer of cellular tissue, will be found a muscle called the TRIANGULARIS STERNI; it arises from the sides of the sternum as high as the third cartilage, from the ensiform cartilage, and sternal extremities of the lower three or four costal cartilages, and is inserted by fleshy digitations into the cartilages of the third, fourth, fifth, and sixth ribs, and often into that of the second. This muscle varies frequently in its extent and points of attachment.

The INTERNAL MAMMARY ARTERY, detached by this section from its connection with the subclavian, descends upon each side of, and about half an inch from the sternum; it gives off the *anterior intercostals*, which turn outward in the upper five or six intercostal spaces to inosculate with the aortic intercostals, and furnish the branches which perforate the intercostal muscles, close to the sternum, to be distributed to the pectoralis major muscle and the integument of the thorax. A small branch, the *comes nervi phrenici*, given off as soon as the artery enters the chest, and which descends to the diaphragm with the phrenic nerve, will have been cut off and left behind. Some small twigs, called *mediastinal* and *pericardiac*, will also be found. At the interval between the sixth and seventh ribs, it gives off a large branch called the *musculo-phrenic*, which winds along the attachment of the diaphragm to the ribs, and supplies the lower intercostal spaces. The termination of the internal mammary is sometimes called the *superior epigastric artery;* it passes downward between the rectus muscle and its sheath, and inosculates with the epigastric branch of the external iliac. This artery occasionally gives off on both sides, from near its origin, a good-sized branch, which might be called an *internal thoracic artery;* it traverses the ribs near their middle, to the fifth intercostal space, where it becomes an intercostal artery.

LIGAMENTS OF THE STERNUM, AND COSTAL CARTILAGES.

The two bones of the sternum are connected by an intervening fibro-cartilage. Upon the anterior and posterior surfaces their union is strengthened by longitudinal fibres, which blend with other similar fibres radiating from the costal cartilages, and, in front, with the sternal tendinous origins of the pectoral muscles. The anterior aspect of the sternum is therefore rough and fibrous, while the posterior is comparatively smooth.

The cartilages of the ribs, received into the lateral fossæ of the sternum, are each provided with a synovial membrane, and held in place by radiating ligamentous fibres, anteriorly and posteriorly, which blend with those of the opposite side, and, in front, with the tendinous origins of the pectoral muscles.

The costal cartilages are held in connection with the anterior extremities of the ribs only by the periosteum.

The cartilages of the sixth, seventh, and eighth ribs articulate by their lower borders with the upper borders of the cartilages next below. The articulation is lined with a synovial membrane, and the connection is maintained by ligamentous fibres. The cartilage of the seventh, and sometimes also of the sixth rib, is attached to the ensiform cartilage by a band of variable size, called the *costo-xiphoid ligament*.

ANTERIOR MEDIASTINUM.

The space left between the two pulmonary cavities is called the MEDIASTINUM; it extends from the summit of the chest to the diaphragm, and from the sternum to the spine. Although there is but one mediastinum, the terms anterior and posterior have been applied respectively to the portions in front of and behind the pericardium.

The ANTERIOR MEDIASTINUM contains a quantity of loose cellular tissue, the remains of the thymus gland, the pericardium and heart, arch of the aorta, superior vena cava, with the right and left innominate veins, bifurcation of the trachea, pulmonary veins and arteries, and phrenic nerves.

The THYMUS GLAND, in the adult subject, consists only of a small quantity of cellulo-adipose tissue; sometimes no trace of it is to be found. In children under two years of age it is an organ of considerable size, reaching from just below the thyroid body, half-way to the diaphragm.

The PHRENIC NERVES, the upper parts of which will have been observed in the dissection of the neck (p. 56), where they arise by filaments from the third, fourth, and fifth cervical nerves, and descend upon the scalenus anticus muscle to enter the chest, will here be seen passing through the anterior mediastinum upon the sides of the pericardium, between it and the pleura. Their course can be traced without dissection, descending to the diaphragm, upon which they ramify beneath the pleura, the two nerves anastomosing on the under surface of the muscle by filaments, which pass through its fibres.

The VENÆ INNOMINATÆ are found in the upper part of the mediastinum. They are formed on each side by the subclavian and internal jugular veins; the union of the two venæ innominatæ constitutes the superior vena cava. Sometimes the innominate veins are not united into one, but descend separately to the heart, where both have distinct openings in the right auricle (p. 55).

The SUPERIOR VENA CAVA is about three inches in length; it passes downward, piercing the pericardium, to enter the upper part of the right auricle of the heart. Before entering the pericardium, it receives the vena azygos major.

If the subject being dissected has been injected from the aorta, the pericardium will have been laid open, in order to perform that operation.

The PERICARDIUM is the sac containing the heart; it consists of a fibrous external and a serous internal layer. The fibrous layer is attached to the great vessels of the heart above, and to the diaphragm below. The serous layer invests the heart, and is then reflected from it to the fibrous layer, thus forming a *shut sac*, as a serous membrane always does, with the heart in reality lying outside of it. The student has only to imagine this sac a globular one to see that the heart, by making a protrusion into its cavity, will not only be covered by a serous surface, but lie in contact with another which is free and external to it.

The HEART occupies an oblique position in the chest, its apex pointing to the space between the fifth and sixth ribs, two or three inches from the sternum, and its base toward the right shoulder. It should be observed that the apex is formed by the left ventricle, as, when the organ is opened and emptied of blood, this is not so apparent. The under side of the heart is flattened, and rests upon the diaphragm; its upper surface is rounded. It may be remarked that the terms "right" and "left" ventricle might well be dispensed with, and the terms "anterior" and "posterior" substituted in their place. Monro states that the habit of describing the two sides as right and left arose from the fact that the earlier dissections were made upon animals, in whom the position of the ventricles differs from that in man, and is in fact right and left.

By lifting the heart, and pulling it to either side, the vessels emanating from or entering it will be made apparent, and are to be examined both within the pericardium and after their exit from it. On the outside a little dissection will be necessary to separate them from one another, and from the bronchial glands and cellular tissue which surround them; part of the pericardium must be cut away.

The superior vena cava will be seen entering the pericardium and joining the upper part of the right auricle.

The INFERIOR VENA CAVA enters the inferior part of the right auricle as soon as it passes through the diaphragm and pericardium, holding the root of the heart downward, and giving the organ the oblique position peculiar to it.

The PULMONARY VEINS are four in number, two for each side; those upon the right side are longer than the others, and emerge from the right lung in front of the pulmonary artery; they pass beneath the inferior vena cava and enter the left auricle. The two left pulmonary veins reach the same cavity after a shorter course, passing in front of the descending aorta. These vessels, though carrying arterial blood, are called veins, because, like veins, they bring the blood *to* the heart.

The PULMONARY ARTERY ascending from the right ventricle between the two auricles, overlies and partially conceals the aorta; at about two inches from its commencement it divides into the right and left pulmonary arteries; at the point of division the remains of the *ductus arteriosus* will be found as a fibrous cord extending from this vessel to the aorta. The left pulmonary artery passes under the arch of the aorta, and, emerging from the pericardium, enters the left lung in front and a little above the left primary bronchus. The right artery passes in front of the descending aorta, and enters the right lung in front and a little below the right primary bronchus. These vessels, though carrying venous blood, are called arteries, because, like arteries, they carry the blood *away* from the heart.

The ARCH OF THE AORTA commences at the anterior part of the left ventricle; emerging between the auricles and behind the pulmonary artery, it ascends, turning gradually to the left, and, passing through the pericardium, curves over the left primary bronchus, giving off the large vessels for the head and arms, which will be studied in connection with the neck (p. 56); it then passes downward into the posterior mediastinum to become the descending aorta. The portion just described is called the arch, from the curved direction which it takes, and is divided into an ascending, transverse, and descending portion. The *ascending* is the part within the pericardium; the *transverse* that from which the great vessels originate; and the *descending* that portion intervening between the last of these vessels and the lower border of the third dorsal vertebra, where the thoracic aorta begins.

The *bifurcation of the trachea* is the division of that air-tube into the right and left primary bronchi. These divisions retain the structure of the trachea, and are made up of similar though smaller and less perfect cartilaginous rings. The *right* bronchus is about an inch in length, and enters the lung above the right pulmonary artery. The *left* bronchus is about two inches in length, and is smaller than the right; it passes obliquely downward under the arch of the aorta, and enters the lung below the left pulmonary artery. The bronchus, with the pulmonary artery and vein, constitute what is called the *root of the lung*.

The BRONCHIAL GLANDS, often numerous and of large size, frequently surround the bifurcation of the trachea. In the adult these are sometimes quite black from carbonaceous deposit, and in scrofulous subjects often contain softened tuberculous or cretaceous matter.

HEART.

The heart should now be removed for further examination by dividing its vessels at such a point as not to mutilate the auricles. The aorta must be cut across near the heart.

The HEART is a muscular organ divided by septa into two halves, right and left; each half consisting of two hollow portions, an auricle and ventricle, and each ventricle, conical in shape, being surmounted by its own auricle; the two ventricles form the bulk of the viscus. The auricles are quadrangular in shape, with a constricted part in front called the *appendix*, and have much thinner walls than the ventricles. The appendices auriculæ project forward with their indented margins, the left being the longest, and nearly meet each other in front of the great vessels. Externally each cavity is defined by a well-marked furrow of separation, containing the ramifications of the vessels destined to the proper nourishment of the heart; that between the auricles and ventricles amounts almost to a constriction of the organ. A quantity of fat fills up the sulcus that would otherwise be formed.

The anterior surface of the heart may be distinguished from the posterior by the appendices of the auricle, which meet in front and not behind, and by the pulmonary artery which lies in front of the aorta. The left ventricle is the thickest and forms the apex of the viscus, a fact which is not always apparent if the heart has become flaccid and shapeless.

Before opening the heart the vessels on its surface are to be dissected. They are the two coronary arteries and the coronary vein.

The CORONARY ARTERIES, two in number, are the first branches of the aorta; they emerge on the sides of the pulmonary artery, and are named right and left. The *right coronary* appears on the right side of the pulmonary artery and winds round between the right auricle and ventricle to the posterior aspect of the heart, where it anastomoses with the left coronary artery, which has followed a similar course on the left side. A branch from the right coronary descends posteriorly in the sulcus between the ventricles. The *left coronary* passes behind the pulmonary artery to emerge on the left side of that vessel and winds round between the left ventricle and auricle to the back of the heart, where it anastomoses with the right coronary artery. The left coronary sends a branch downward to the apex of the heart in the anterior sulcus between the ventricles.

The *anterior* and *posterior cardiac veins* accompany these arteries and terminate in the GREAT CARDIAC, or CORONARY VEIN; this occupies the sulcus between the right auricle and ventricle posteriorly, and winding round to the front terminates in a dilated ending, called the CORONARY SINUS; its termination in the sinus is marked by two valves, and the sinus consists of the portion intervening between these two valves and the coronary valve where it opens into the right auricle.

The nerves of the heart are furnished by the *superficial cardiac plexus*, which surrounds the origins of the aorta and pulmonary artery. The *deep cardiac plexus* lies between the trachea and the arch of the aorta. These plexuses are derived from the sympathetic and pneumogastric nerves, and in them terminate the cervical cardiac branches (p. 53).

> The right auricle is opened by introducing the point of the scissors into the superior vena cava, and cutting toward, and nearly to, the inferior vena cava; from the middle of this incision another is made to the tip of the appendix.

In the RIGHT AURICLE the following parts are to be noticed:—

The *endocardium* is the smooth transparent lining membrane, common to all the cavities of the heart, and continuous with the inner coat of the vessels. When it passes from an auricle to a ventricle, or from a ventricle to an artery, it forms duplicatures, or valves, in which fibrous tissue is inclosed.

The *Eustachian valve* is situated between the lower cava and the auriculo-ventricular opening; it is a semilunar fold, sometimes rudimentary and sometimes developed into a well-formed valve with a reticulated margin. During fœtal life this structure serves to direct the blood from the inferior cava toward the foramen ovale.

The *coronary valve* is occasionally connected with the preceding; this also is a semilunar fold stretching across the orifice of the coronary sinus, which enters the auricle just below the inferior vena cava.

The *foramina Thebesii* are the apertures of minute veins found in various parts of the cavity, and which, coming from the muscular structure of the heart, pour their contents directly into the auricle.

The *fossa ovalis* is a rounded depression situated on the septum between the auricles, characterized by a well-defined margin, and called the *annulus ovalis;* it is the remains of the foramen ovale of fœtal life. Not unfrequently the annulus forms a sort of valve upon one side of the fossa, beneath which a probe may be insinuated and passed into the left auricle, the foramen remaining imperfectly closed; this may be consistent with an undisturbed condition of the circulation.

The *musculi pectinati* are parallel muscular columns, symmetrically arranged, and chiefly confined to the parietes of the auricular appendix.

The *auriculo-ventricular orifice* consists of a fibrous ring, with which are connected the folds of membrane constituting the tricuspid valve.

The right ventricle is opened by introducing one blade of the scissors through the auriculo-ventricular orifice, and incising its wall along the outer edge, nearly to the apex of the heart; from the ventricle pass the scissors into the pulmonary artery, and make an incision, which, commencing at the termination of the former, shall pass upward, parallel to the ventricular septum, dividing the pulmonary artery. The V-shaped flap thus made, when lifted, will expose the ventricle.

The RIGHT VENTRICLE is remarkable for its fleshy bands and the generally irregular character of the surface of its cavity; near the aperture of the pulmonary artery, however, its walls become comparatively smooth.

The *columnæ carneæ* are the muscular columns, interlacing in all directions, which give the ventricle its characteristic appearance.

The *chordæ tendineæ* are small tendinous bands attached to certain of the columnæ carneæ, and extending from them to the free edge of the tricuspid valve; they interlace with each other, and several of them converge to one column for attachment.

The *tricuspid valve* consists of three or more folds of the lining membrane of the heart, strengthened by fibrous tissue, and attached by their base to the auriculo-ventricular orifice, and by their free edge, which is usually a little thicker than elsewhere, to the chordæ tendineæ. This valve obstructs the regurgitation of the blood from the ventricle to the auricle, the chordæ tendineæ preventing its flaps from being pressed through into the auricle.

The *infundibulum*, or *conus arteriosus*, is that dilated portion of the ventricle from which the pulmonary artery arises; it is, as it were, separated from the rest of the ventricle by a sort of constriction; it has fewer columnæ carneæ than other parts of the cavity.

The lining membrane, at the commencement of the pulmonary artery, forms three crescentic folds, called *sigmoid*, or *semilunar valves*. Attached by their base, they are free along their concave margin, and look upward in the course of the vessel. The margin of these valves is often perforated by small openings, and each valve contains in the centre of its concavity a little fibrous nodule, called the *corpus Arantii;* this is sometimes directly at the edge, at other times a little distant from it. Behind each of these valves the pulmonary artery forms dilatations like those similarly placed in the aorta, but not nearly so well marked.

The left auricle may be opened by a transverse incision along its ventricular border.

The LEFT AURICLE is smaller than the right, but has, however, somewhat thicker walls. It has four openings for the pulmonary veins, two upon each side; occasionally, two of them coalesce. The septum auriculæ has, upon this side, no trace of the fossa ovalis, except when an opening exists, and then a small valvular fold may be observed. The musculi pectinati are found only in the appendix. The auriculo-ventricular orifice is smaller than that of the right side.

To expose the left ventricle, make a short incision into it with a knife close to the septum; pass the finger through this into the aorta, and upon it, as a director, divide the ventricle and aorta with the scissors, keeping close to the septum, passing between the two appendices auriculæ, and holding aside the pulmonary artery, so as not to injure that vessel in making the section.

The LEFT VENTRICLE is more conical in shape than the right, and its walls are twice as thick and much more muscular; its surface is irregular from the columnæ carneæ, but near where the aorta arises the walls are smooth.

The *mitral* valve will have been left undivided if the ventricle is opened as has been directed. It consists of two folds of the lining membrane, attached to the auriculo-ventricular orifice, and connected to the columnæ carneæ by chordæ tendineæ in a manner similar to that described as existing upon the right side. The chordæ tendineæ converge to be inserted in two distinct bundles into the columnæ carneæ; they are stronger, but less numerous, than those of the right ventricle, and the same may be said of the columnæ carneæ.[1]

The *septum ventriculorum* seems to form part of the left rather than of the right ventricle, being concave on its left and convex on its right side.

The *aortic semilunar valves* are found at the commencement of the aorta, nearly on a level with the mitral valve. They are stronger, though similar in number and in general shape to those of the pulmonary artery; the corpora Arantii are more developed, and the dilatations of the vessel behind each segment are much more pronounced; they are here called the *aortic sinuses*, or *sinuses of Valsalva*.

The orifices marking the origin of the *anterior* and *posterior coronary arteries* will be observed just behind the semilunar valves. Occasionally, these arteries arise from a common trunk, and their number is sometimes increased to three. Their course has been described (p. 116).

The student should at some time avail himself of an opportunity to examine the interior of the heart, prepared after the following method. Having washed out the blood and fibrin contained in its cavities, they are to be distended, by filling them from a syringe, with undiluted alcohol, and confining it there by tying all the vessels; then immersing the whole organ in a jar filled with alcohol, in a few

---

[1] The letters L. M. (so familiar in another connection) will give the student a mnemonic key to the side to which the mitral valve belongs. (L. M., left, mitral.)

days it will become so stiffened and hardened that it may be opened and the interior examined, by windows cut into the cavities, without its walls collapsing; the valves will be preserved so nearly in their natural condition as to convey a much clearer idea of their character and relations than can be otherwise obtained.

DISSECTION IV.

POSTERIOR MEDIASTINUM.

To examine the posterior mediastinum, the left lung should be lifted from its pleural cavity and turned over to the opposite side, where it should be confined, or held resting in the space that was occupied by the heart. It may be desirable, in order better to get at the parts to be dissected, to remove the anterior halves of the ribs on one side; this can be accomplished by the saw, or by bone forceps.

The POSTERIOR MEDIASTINUM is bounded in front by the posterior surface of the pericardium, behind by the vertebral column, and on each side by the pleura. It contains the thoracic aorta and vena azygos, the œsophagus and thoracic duct, and the pneumogastric and sympathetic nerves.

The continuity of the pleura, as it is reflected from the base of the lung to the vertebræ, ribs, and diaphragm, will now be well seen, and the present, perhaps, affords the best opportunity of appreciating the manner in which one part of a serous membrane is attached to the parietes of a cavity, and the other to the organ contained in it; the organ being in reality outside of it, and merely pushing inward the part covering it.

The left pleura is to be carefully dissected from the subjacent parts, the whole length of the thorax. This will expose the contents of the mediastinum.

The PNEUMOGASTRIC NERVES will be found lying, the left upon the anterior surface of the œsophagus; the right upon the posterior surface. They give off numerous filaments to the œsophagus as they descend upon it to the stomach, where they terminate in *gastric* branches. They also give off *pulmonary* and *cardiac* branches, and form a large *plexus* behind the root of the lung, with which are connected some filaments from the gangliated cord of the sympathetic. In the upper part of the thorax they send off the *recurrent laryngeal nerves*, which pass upward beside the trachea to the larynx; upon the left side this nerve curves round the arch of the aorta; on the right it curves round the subclavian artery (p. 56).

The THORACIC AORTA commences at the lower border of the third dorsal vertebra; it lies at first upon the left side of the vertebral column, and then inclines inward to the median line; as it passes through the diaphragm it rests upon the fronts of the vertebræ. It gives off the following branches, viz:—

    Bronchial,
    Œsophageal,
    Intercostal.

The *bronchial arteries*, two or more in number, and very irregular in their origin, are usually found on the anterior aspect of the aorta, just as the arch ceases to make its curve; they are of considerable size, and pass immediately to the primary bronchi, upon which they ramify in a tortuous manner, giving small twigs to the œsophagus and pericardium, and terminating in the parenchyma of the lungs, of which they are the nutrient vessels.

The *œsophageal arteries* vary in number, and are small branches arising from the front of the aorta, and distributed upon the œsophagus.

The *intercostal arteries*, nine in number, upon each side, are given off from the posterior aspect of the aorta. The right intercostals are the longest, the position of the aorta obliging them to arch over the bodies of the vertebræ. They supply each intercostal space except the two upper, which are furnished by a branch from the subclavian (p. 59). Each artery is accompanied by a nerve and vein, the former being the anterior branch of the spinal nerves, and the latter a branch of the vena azygos. The artery occupies the upper part of the intercostal space, lying in the groove of the lower border of the rib, between the two layers formed by the external and internal intercostal muscles; it passes forward to inosculate with the anterior intercostal branch of the internal mammary, giving off at various parts of its course *external* branches, which perforate the intercostal space to go to the muscles and integument of the back and thorax. From one of these branches in the dorsal region a *spinal* twig goes to the interior of the vertebral canal.

The lungs should now be removed, by dividing the trachea just above its bifurcation.

The ŒSOPHAGUS is a hollow muscular tube, extending from the pharynx to the stomach. The cervical portion is described at p. 62. The thoracic portion enters the chest on the left of the median line, passes beneath the arch of the aorta, to continue on the right side of that trunk to the lower part of the chest, where it again inclines to the left, over the aorta, and passes through the œsophageal opening of the diaphragm.

The œsophagus has a muscular and a mucous coat; the muscular coat is made up of longitudinal and circular fibres, and is connected with the mucous by an intervening layer of cellular tissue; the mucous coat is of a pale color, and moves freely upon the muscular, the contraction of which throws it into longitudinal folds.

The SYMPATHETIC NERVE consists of two portions. The first is the *prevertebral portion*, made up from the cardiac nerves, descending from the cervical ganglia (p. 53), the branches of which, uniting with filaments from the recurrent laryngeal and pneumogastric nerves, form the *cardiac* and *pulmonary plexuses*, distributed over the origin of the great vessels of the heart, and to the heart itself, the root of the lungs, and the trachea. These can only be satisfactorily studied by special dissections.

The second or *vertebral portion* of the sympathetic, consists of a chain of twelve connected ganglia, situated near the heads of the ribs, covered in by the pleura, and continuous with those of the neck and abdomen. The upper ganglion is the largest, and the two lower are anterior to the line of the others. Each ganglion furnishes a branch to the intercostal nerves, and from the upper six, small branches are sent to the aorta and mediastinum. Branches from the sixth, seventh, eighth, and ninth, unite to form the *great splanchnic nerve*, which passes through the diaphragm, by the side of its crus, to join the semilunar ganglion. Branches from the tenth and eleventh ganglia form the *lesser splanchnic nerve*, which, piercing the diaphragm, goes to join the renal and cœliac plexuses. A branch from the twelfth ganglion, occasionally communicating with the preceding nerve, also pierces the diaphragm, and joins the renal and cœliac plexuses, under the name of the *third* or *renal splanchnic nerve*.

The INTERCOSTAL NERVES, twelve upon each side, are the anterior branches of the spinal nerves of the dorsal region. They pass forward between the two muscular layers of the intercostal space, perforating the external muscle anteriorly, to be distributed to the integument of the front of the thorax. Each nerve receives a short branch from the ganglionic trunk of the sympathetic. The first intercostal nerve sends a large branch to join the brachial plexus; the second gives off the intercosto-humeral branch (p. 108), and the twelfth a branch to the first lumbar nerve, to assist in forming the lumbar plexus.

By removing a portion of the œsophagus, the azygos veins will be brought into view; they are two in number, and are named major and minor.

The AZYGOS MAJOR VEIN, commencing by branches communicating with the right lumbar and renal veins, and sometimes also with the inferior vena cava, passes through the aortic opening of the diaphragm, and ascends upon the bodies of the vertebræ at the right side of the thoracic aorta, receiving the right intercostal veins in its course. Opposite the third intercostal space it arches forward above the root of the right lung, and enters the superior cava, just before that vessel penetrates the pericardium.

The AZYGOS MINOR VEIN commences on the left side from the lumbar, or renal veins, passes into the thorax with the aorta, or beneath the border of the diaphragm, ascends on the left side of the vertebral column, and at about the seventh or eighth dorsal vertebra, crosses beneath the aorta and thoracic duct, to enter the vena azygos major. It receives the lower intercostal veins of the left side. The superior intercostal veins enter a trunk which joins with the left vena innominata, or with the azygos minor.

<small>The thoracic duct is difficult to demonstrate or isolate from the surrounding tissues, which it resembles in color; by making a snip with the scissors into the duct, then inserting the blow-pipe, and inflating it, it will become distended, and its course made apparent.</small>

The THORACIC DUCT commences in an enlargement, to which the lymphatics of the abdomen converge, called the receptaculum chyli; this lies beside the right crus of the diaphragm, between the aorta and vena cava. The duct, about the size of a wheat straw, ascends between the aorta and vena azygos major, crosses the vertebral column at the second dorsal vertebra, and continuing upward along the left side of the œsophagus, enters the left subclavian vein near its junction with the internal jugular. A small duct, called the *ductus lymphaticus dexter*, being the terminal duct of the lymphatics of the head, neck, and portions of the right side of the upper part of the body, enters the right subclavian at its junction with the right internal jugular vein (p. 55). The thoracic duct is sometimes double, either in the whole or part of its course. It occasionally empties into the vena azygos major, that being its normal destination in some mammalia. The duct has been found also on the left side of the aorta.

The intercostal spaces are filled by two muscular layers, called external and internal intercostal muscles.

The EXTERNAL INTERCOSTAL MUSCLES are eleven in number on each side; they arise from the outer lip of the lower border of the rib, and are inserted into the corresponding part of the rib next below; the fibres run forward and downward, and extend from the tubercle of each rib nearly to its cartilage.

The INTERNAL INTERCOSTAL MUSCLES, also eleven in number on each side, arise from the inner lip of the lower border of the rib, and are inserted into the upper border of the rib next below; they extend from the sternum to the angles of the ribs, and their fibres are directed backward and downward. The internal muscular layer is covered by the pleura, and the intercostal vessels and nerves ramify between the two muscles.

The INFRA-COSTALES MUSCLES are bundles of muscular fibres on the inner surface of the ribs, having the same direction as the internal intercostals: they stretch across two or three spaces, and vary in size and number. They are most constant on the lower ribs.

LUNGS.

Dissecting-room subjects rarely afford a good opportunity for the examination of the lungs; riddled with tubercular cavities, bound down by old pleuritic adhesions, they are apt to be mutilated in their removal, or rendered puzzling to the student, by their unnatural condition.

If uninjured, the lungs should be inflated for their examination.

The LUNG is conical in shape, and presents a rounded *apex* which extends above the first rib, and a concave *base*, or diaphragmatic surface, the concavity being greatest in the right lung, owing to the position of the liver; the sharp border of this surface penetrating the space between the diaphragm and ribs posteriorly, makes the posterior longitudinal measurement of the lung greater than the anterior. The sides of the lung, with the exception of the mediastinal, present a smooth convex surface, covered with the pleura, and conforming to the shape of the thoracic cavity. The mediastinal surface is concave, the position of the heart making the concavity of the left lung the greatest. The anterior border is sharp, and the posterior rounded. Each lung is divided into an *upper* and *lower lobe* by a deep fissure, the upper lobe of the right side being subdivided by a more shallow fissure, thus making a *third*, or *middle lobe*. The lobes are sometimes multiplied, and offer great variety in shape. The lungs vary in color, according to age, or their more or less healthy state; usually, they are of a grayish tint, mottled with blackish spots. The surface is figured with irregular polygonal outlines, indicating the lobules of which they are made up, and these lobules are subdivided by still smaller lines which are the walls of the cells that compose them; the lobules are best seen in infant's lungs, or in those of very young persons. When cut into, the lungs are found to be of a spongy texture, and if in a natural state, upon pressure, the air may be felt escaping in fine bubbles from the air-cells, giving the sensation called *crepitation*.

The lung is made up of the various structures which enter it at its root. The bronchi may be traced by a director and scissors, dividing into principal trunks for each lobe, and then subdividing, until lost by their extreme tenuity; the cartilaginous rings which were found at their commencement, becoming less and less apparent.

At the divided root of the lungs, the bronchus is posterior, and the pulmonary veins anterior, the pulmonary artery being between the two. In the direction from above downward, the position on the right side is—bronchus, pulmonary artery, and pulmonary veins; but, on the left side, it is changed to the order—artery, bronchus, and veins.

DISSECTION V.

THE BACK AND POSTERIOR CERVICAL REGION.

The dissection of the back comes next in order, but if the other members of the class are not ready to turn the subject over, the dissection of the arm may be resumed, the description of which will be found in Dissection VII.

The subject must be turned upon its face, and rest upon blocks as before; the head should hang over the end of the table, and the arms over its sides; by so doing, the muscles will be put upon the stretch. A longitudinal incision is to be made along the median line, from the occiput to the sacrum, and another from just below the middle of the dorsal vertebræ to the acromion of the scapula; the flaps, thus formed, are to be raised and reflected.

The muscles of the back, always difficult to dissect neatly, are often made more so, by the infiltration of fluids which have gravitated to this part of the body, while it remained dependent. The subcutaneous cellular tissue is sometimes so much thickened by this infiltration, that it is difficult to tell when the plane of the muscles is reached. If this condition of things exist, care must be taken not to go through the thin aponeurotic tendons by a too hasty incision.

The dissection of the back is usually made for the benefit of the class in common (with the exception of the cervical region, which belongs more properly to the head), and is generally accomplished by two of the class, the others assisting by reading the description of the parts successively dissected. The student must be prepared to find a great want of conformity between the muscles, as he finds them, and the precise description of their origins and insertions as given in books, for they have little of the distinct arrangement elsewhere to be found. Of many, the fibres are so short, or so incompletely separated, that their isolation is extremely difficult. "The deeper ones," says John Bell, " might fairly be reckoned as one muscle, since they are one in place and in office, but which the anatomist may separate into an infinite number, with various and perplexing names, an opportunity which anatomists have been careful not to lose."

The trapezius and the latissimus dorsi muscles form the superficial layer, and together cover the whole region of the back.

The TRAPEZIUS MUSCLE arises by a thin aponeurotic tendon (easily divided and injured if much care is not taken), from the occipital protuberance and adjacent part of the superior curved line of the occipital bone, from the ligamentum nuchæ, and from the spinous processes, and supraspinous ligament of the last cervical vertebra, and of a variable number (six to twelve) of those of the dorsal region; from this extended origin the fibres converge, so as to give the muscle a triangular shape, and are inserted into the outer third of the clavicle, the acromion process, and the spine of the scapula. The spinal accessory nerve pierces the anterior border of this muscle and is distributed to its fibres.

The LIGAMENTUM NUCHÆ is a strong layer of elastic fibrous tissue, extending from the spine of the occiput to that of the seventh cervical vertebra; it is a rudimentary development of the elastic band which serves to sustain the weight of the head in the Ruminantia.

The LATISSIMUS DORSI MUSCLE arises from the spinous

processes, and supra-spinous ligament of from four to eight of the dorsal, those of all the lumbar, and two of the sacral vertebræ, from the posterior third of the crest of the ilium, and from three or four of the lower ribs, by serrations which indigitate with similar processes of the external oblique muscle of the abdomen; the fibres pass upward and forward to be inserted by a strong flat tendon into the floor of the bicipital groove of the humerus. In their course upward they overlap the inferior angle of the scapula, and beneath them, at this point, a synovial bursa may sometimes be found. A distinct fleshy slip is sometimes given off from the lower angle of the scapula, and a muscular band often stretches across the axilla to terminate in either the pectoralis major or the coraco-brachialis muscle.

Cutaneous nerves, branches of the posterior divisions of the spinal nerves, will be observed perforating the spinal tendons of both the last-described muscles.

A triangular space will sometimes be found intervening between the latissimus dorsi and external oblique, only remarkable as having been erroneously supposed to be a point at which intestinal hernia was liable to occur.

The latissimus dorsi is covered in its dorsal region by the trapezius muscle and, at its lower part, by a layer of the *fascia lumborum;* this fascia consists of three layers; the superficial, lying upon the latissimus and blending with its aponeurosis is attached to the two lower ribs, and the spines of the lumbar vertebræ; the middle passes beneath the latissimus, between the erector spinæ and quadratus lumborum, and is attached to the tips of the lumbar transverse processes; the internal passes in front of the quadratus lumborum, and is attached to the bases of the lumbar transverse processes. These three layers constitute the posterior origin of the transversalis abdominis muscle.

> The trapezius and latissimus are now to be divided through the middle of their muscular portions and reflected. In doing this, care must be taken not to push too far the separation between them and the muscles beneath at their spinal attachments, inasmuch as their tendons are united and confounded with those of the next layer. In removing the trapezius, branches of the supra-scapular artery will be divided.

The RHOMBOIDEUS MINOR MUSCLE is a narrow band arising from the spinous processes, and supra-spinous ligament of the last cervical and first dorsal vertebræ, and inserted into that part of the border of the scapula opposite the triangular space by which its spine commences.

The RHOMBOIDEUS MAJOR MUSCLE, double the width of the preceding, arises from the spinous processes and from the supra-spinous ligament of the upper four or five dorsal vertebræ, and is inserted into the posterior border of the scapula, below the spine; it is separated from the rhomboideus minor by a slight cellular interspace.

These two muscles must be carefully removed, in order to expose those next to be described; in so doing twigs of the posterior scapular artery will be found distributed to them, and that artery must be respected, as it passes along the posterior border of the scapula.

The SERRATUS POSTICUS SUPERIOR is a thin muscular plane, arising from the spinous processes of one or two of the last cervical and two or three of the upper dorsal vertebræ; it passes downward and outward, and is inserted by fleshy serrations into the upper borders of the second, third, and fourth ribs.

The SERRATUS POSTICUS INFERIOR, the tendon of which is inseparably connected with the aponeurosis of the latissimus dorsi and the fascia lumborum, arises from the spinous processes of the last two dorsal and first two or three lumbar vertebræ, and passing obliquely upward, is inserted by fleshy serrations into the last four ribs, each successive process extending further outward than the one below.

A thin aponeurotic lamina, called the *vertebral aponeurosis* extends between these two last-described muscles, binding down the deeper muscles, and separating them from the more superficial ones.

By dividing the serratus posticus superior and the subjacent vertebral aponeurosis the splenius muscle will be exposed.

THE SPLENIUS MUSCLE, single in its origin, divides into a cervical and cranial part, known respectively as splenius capitis and splenius colli. It arises from the spinous processes and inter-spinous ligaments of the last two cervical and four or five upper dorsal vertebræ, and, separating into its two divisions, the *splenius capitis* is inserted into the space between the two curved lines of the occipital bone and into the mastoid process of the temporal bone, where it is overlapped by part of the sterno-mastoid muscle; the *splenius colli* is inserted into the posterior tubercles of the transverse processes of three or four of the upper cervical vertebræ.

The splenii muscles of the two sides do not meet along the median line, but leave a space between them filled with dense areolar tissue, the removal of which displays the complexus muscle beneath.

By dividing the splenius muscle the whole of the levator anguli scapulæ will be exposed.

The LEVATOR ANGULI SCAPULÆ arises by distinct slips from the posterior tubercles of from three to five of the superior cervical vertebræ, between the insertions of the scalenus posticus and splenius muscles, which latter, with the sterno-mastoid muscle, overlap a portion of it, and is inserted by a fleshy tendon into the superior angle of the scapula. It is sometimes split into several distinct muscles, the divisions of its origin continuing down to its insertion.

The OMO-HYOID MUSCLE, connecting the scapula and hyoid bone, consists of two bellies, an anterior and a posterior. The anterior is described at p. 40. The posterior belly is now seen. This portion arises from the upper border of the scapula, near the supra-scapular notch, and from the ligament which converts that notch into a foramen; it is thin and ribbon-like, and terminates beneath the sterno-mastoid muscle in a tendon which separates the muscle into its two halves, and which plays through a loop formed by the deep cervical fascia. This loop holds down the tendon so that the portion of the muscle just described forms an obtuse angle with that part inserted into the os hyoides.

The *supra-scapular* nerve (p. 103) passes through the supra-scapular notch beneath this muscle. The *supra-scapular artery* (p. 58) also passes beneath this muscle, to the supra-spinous fossa. The *posterior scapular artery* (p. 59) passes under the levator anguli scapulæ, and turns downward along the base of the scapula, beneath the rhomboid muscles, to supply the two surfaces of that bone.

In the interstices of the muscles which remain, and in those which are to be made in separating them from each other, numerous small arteries will be seen; in the lumbar region, they are *posterior* branches from the *lumbar* arteries; in the dorsal region, *posterior* branches from the *intercostal* arteries; and in the neck, *posterior* branches from the *vertebral* arteries, and from the *profunda cervicis* and *super-*

ficialis cervicis, offsets from the subclavian artery. The *occipital artery*, a branch of the external carotid (p. 45), will also be seen emerging from beneath the tendons of the sterno-mastoid, trachelo-mastoid, and splenius muscles, to ramify upon the occipital bone; this artery sends off a branch called the *princeps cervicis*, which passes downward between the complexus and semi-spinalis colli muscles, to inosculate with the profunda cervicis.

The *posterior* branches of the *sacral, lumbar, dorsal*, and *cervical* nerves, will also be observed; they are mostly small filaments accompanying the arteries, and supplying the muscles and skin. The posterior branches of the first, second, and third cervical nerves are however larger than the others, and form a plexus upon the muscles of the sub-occipital region, called the *posterior cervical plexus*. The *occipitalis major nerve*, from the second cervical, is a branch of considerable size, which passes upward, in company with the occipital artery, and ramifies between the integument, which it supplies, and the posterior belly of the occipito-frontalis muscle; it sends numerous branches to the muscles of the back of the neck.

Under the title of ERECTOR SPINÆ are included a number of muscles of different length, extending from the sacrum to the upper part of the neck; small and pointed over the sacrum, in the lumbar region the erector constitutes a single, inseparable mass; in the dorsal region its bulk gradually lessens, and in the neck it consists only of slender prolongations. A strong and lustrous tendinous expansion covers the sacral and lumbar portion; and from this, from the posterior surface of the sacrum, and from the whole length of each transverse process of the lumbar vertebræ, and the layer of the fascia lumborum external to these, the erector spinæ takes its origin. At the level of the last rib, the muscle begins to separate into two portions, the external being called the sacro-lumbalis, and the internal the longissimus dorsi.

The *sacro-lumbalis* muscle is inserted by separate tendons into the angles of the six lower ribs; here the muscle is reinforced by muscular fasciculi arising from the upper margin of all the ribs, internally to the preceding insertions, and through them the sacro-lumbalis is continued to the higher ribs, and to the transverse processes of three or four lower cervical vertebræ. There is no separation between these accessory fasciculi and the bulk of the sacro-lumbalis; but those derived from the six or eight lower ribs and inserted into the upper ribs, are often described separately, as the *musculus accessorius ad sacro-lumbalem;* and those from the four or five upper ribs which are inserted into the cervical transverse processes, as the *cervicalis ascendens* muscle.

The *longissimus dorsi* muscle is inserted into the transverse processes of all the dorsal vertebræ, and to that portion of from seven to eleven ribs, which is situated within their angles;—it is continued upward into the neck by a slender accessory portion, often described as a distinct muscle, under the name of the *transversalis colli,* and which, arising from the tips of the transverse processes of the four upper dorsal and the seventh cervical vertebræ, is inserted into the transverse processes of about four cervical vertebræ, above the preceding, blending with the cervicalis ascendens and trachelo-mastoid muscles.

The *trachelo-mastoid* muscle is the continuation of the longissimus dorsi to the head. It arises from the articular processes of three or four lower cervical vertebræ, on the inner side of, and inseparable from, the transversalis colli, and is inserted into the posterior part of the mastoid process, beneath the splenius and sterno-mastoid muscles.

The *spinalis dorsi* is often described as a part of the longissimus dorsi, and it can only be artificially separated from it. It arises by separate tendons from the spinous processes of the first two lumbar and last dorsal vertebræ, and is inserted into from four to eight of the spinous processes of the upper dorsal vertebræ. The muscles of the two sides form a long ellipse.

The COMPLEXUS MUSCLE, with the splenius, forms the bulk of the back of the neck. It arises from the transverse processes of four upper dorsal, and from the transverse and articular processes of four lower cervical vertebræ, and is inserted into the occipital bone between the curved lines. Upon its inner border, a large fasciculus, consisting of two bellies with an intervening tendon, from which peculiarity it has been named *biventer cervicis,* separates itself from the principal mass of the complexus.

The occipitalis major nerve is transmitted to the surface through the complexus and trapezius muscles, near their cranial attachment. As soon as the nerve is free from the muscles, it receives a cutaneous offset from the third cervical nerve. The occipital artery rests upon the upper end of the complexus, and beneath it the branch of that artery called the princeps cervicis anastomoses with the profunda cervicis.

The *posterior belly* of the *occipito-frontalis* muscle (p. 15) will be seen during this dissection. It is a thin, flat plane of muscular fibres arising from the outer part of the superior curved line of the occipital bone, and inserted into the epicranial aponeurosis, by which it is connected with its frontal portion.

The erector spinæ, with its accessories and the complexus, are now to be removed by dividing transversely the tendon of the former, close to the sacrum, raising it from the inner side and turning it outward. The muscles which remain to be examined, with the exception of the sub-occipital group, will be left in a very ragged condition, and obscured by the remains of the numerous tendons and fasciculi divided in the removal of the dissected muscles. It is difficult to make a neat preparation of them.

The SEMI-SPINALIS MUSCLE consists of a thin and narrow stratum of short muscular bellies with longer tendons, which stretches from the second cervical vertebra to the lower part of the dorsal region; each bundle, arising from a transverse process, is inserted into a spinous process, the fibres being directed downward and outward. The upper bundles are larger than the lower, and the number of them varies in different subjects. Although continuous, the upper four of these muscular bundles, those arising from the transverse processes of the four upper dorsal and inserted into the spinous processes of the four upper cervical vertebræ, have been named the *semi-spinalis colli*,—and the lower six, those arising from the transverse processes of the six lower dorsal and inserted into the spinous processes of the four upper dorsal and two lower cervical vertebræ, the *semi-spinalis dorsi*.

The MULTIFIDUS SPINÆ lies to the inner side of and beneath the last-named muscles, and they must be removed in order to see it in its full extent. It reaches from the sacrum to the axis, and consists of a series of muscular slips, filling the vertebral groove at the side of the spinous processes. Each fasciculus arises from a transverse process, and is inserted into the spinous process of the first or second vertebra above; the first slip arises from the transverse process of the third cervical vertebra, and is inserted into the spinous process of the axis; the last slip arises from the back of the sacrum, and is inserted into the spine of the fifth lumbar vertebra.

The LEVATORES COSTARUM are sometimes considered as accessories of the external intercostal muscles; triangular in shape, they arise from the transverse processes of the dorsal vertebræ, and are inserted between the tubercle and the angle of the rib below. The inferior levatores sometimes pass over one rib, to be attached to the second below them. There are twelve of these muscles on each side.

Between the spinous processes of the cervical and lumbar regions may be found a series of small muscles called

INTER-SPINALES. In the neck, where the spines are bifid, they are arranged in pairs; they are wanting between the first two cervical vertebræ, and in the dorsal region they are rudimentary; as their name indicates, their origin is from one spinous process, and their insertion is into that of the next vertebra below it.

Similarly disposed to these last are the INTER-TRANSVERSALES MUSCLES, best marked in the cervical region, where they are arranged in pairs, corresponding to the anterior and posterior tubercles of the transverse processes between which they are arranged; the posterior muscle in the upper inter-transverse space is often wanting. The posterior branches of the spinal nerves emerge between the inter-transversales muscles.

A better defined series of muscles remains to be examined; they are those which communicate to the head its peculiar movements; they are covered in by an aponeurosis of fibrous tissue, which is to be removed; in so doing, the sub-occipital, or first cervical nerve, is to be respected.

The RECTUS CAPITIS POSTICUS MAJOR arises from the spine of the axis, and spreading, fan-like, is inserted into and beneath the inferior curved line of the occipital bone. It diverges from its fellow so as to leave a deep interspace between them.

The RECTUS CAPITIS POSTICUS MINOR is placed internally to the preceding; it arises from the posterior border of the atlas, and is inserted into the occipital bone between the inferior curved line and the foramen magnum.

The OBLIQUUS INFERIOR arises from the spinous process of the axis, externally to the rectus major muscle, and is inserted into the transverse process of the atlas.

The OBLIQUUS SUPERIOR arises from the transverse process of the atlas, passes upward and forward, and is inserted just behind the mastoid process, between the curved lines of the occipital bone.

The SUB-OCCIPITAL NERVE, the posterior division of the first cervical nerve, pierces the ligament between the first cervical vertebra and the occipital bone, and appears in the interval between the recti and obliqui muscles; it is distributed to these muscles, and sends a branch downward to communicate with the second cervical nerve.

The strikingly symmetrical arrangement of the occipital group of muscles cannot but be noticed. The muscles parting from the spine of the axis form a *star* with six points; the inferior points being formed by the semi-spinales colli, the lateral by the obliqui inferiores, and the superior by the recti capitis majores.

SPINAL CORD AND MEMBRANES.

The muscles are to be dissected away from the sides of the vertebral spines. With the chisel and saw their arches are to be divided upon each side close to the articular processes; the bones can only be removed piecemeal and with difficulty. This done, the membranes of the spinal cord will be exposed.

The membranes of the spinal cord are covered externally by veins and by a loose areolar tissue containing fat and, especially at the lower part, a little fluid; they are a continuation of those of the brain, and, like them, consist of dura mater, arachnoid, and pia mater.

The DURA MATER envelops the cord loosely, and sends tubular prolongations along the spinal nerves issuing at the intervertebral foramina; at its lower part these prolongations become longer and lie for some distance within the spinal canal. The dura mater terminates in an impervious fibrous process, which blends with the periosteum covering the back of the coccyx.

The dura mater is to be opened lengthwise with the scissors; this will expose the arachnoid.

The ARACHNOID is a serous membrane enveloping the spinal cord, and reflected upon the internal surface of the dura mater. That portion attached to the dura mater is closely adherent to it, while that in relation to the cord is loose; the interval between the cord and membrane constitutes the *sub-arachnoid space*, and is filled by a *fluid* called the *cerebro-spinal* (p. 80). The arachnoid envelops each spinal nerve and the collection of nerves which terminates the cord.

The loose arachnoid is to be removed and the pia mater will then be exposed.

The PIA MATER is a thin and stout membrane closely investing the spinal cord; it forms a sheath for the spinal nerves and inferiorly is prolonged downward in a slender process called the *filum terminale*, which blends with the terminal prolongation of the dura mater.

On each side of the spinal cord, extending its whole

length, and separating the anterior and posterior roots of the spinal nerves, is a white fibrous band, connected internally with the pia mater, and having about twenty serrations along its free margin which connect it externally with the dura mater; from this peculiarity, and from its supporting the cord, it receives the name of *ligamentum denticulatum*, or *membrana dentata*.

The SPINAL CORD gives off *thirty-one pairs of nerves*, arising by two roots and passing out at the intervertebral foramina; they are divided into groups, which are named cervical (eight pairs), dorsal (twelve pairs), lumbar (five pairs), sacral (five pairs), and coccygeal (one pair); in each group the nerves are equal to the number of vertebræ, except in the cervical, which has eight, and in the coccygeal, which has but one; as the cervical nerves exceed the number of cervical vertebræ, the lowest nerve of each group is consequently below its corresponding vertebra.

The two *roots* which blend to form the spinal nerves are called *anterior* and *posterior*, or *ganglionic* and *a-ganglionic*. The posterior roots are the largest, and are each furnished with a ganglion. As the apertures for the transmission of the nerves are not opposite their points of origin, they get an oblique direction, increasing from above downward; in the lumbar and sacral region their direction is vertical, and the collection of the roots of the nerves around the filum terminale, which constitutes about one-third of the whole length of the cord, is called the *cauda equina*. It is upon these lower nerves that the ganglion of the posterior root may be best observed, as in those nerves given off more nearly opposite their foramina it often lies in the intervertebral canal. The first nerve sometimes wants a posterior root.

In the upper part of the canal, the *spinal* portion of the *spinal accessory nerve* (p. 83) should be sought; it arises by fine filaments from the side of the spinal cord as low down as the sixth cervical nerve, and lies between the membrana dentata and the posterior roots of the spinal nerves, with the upper of which it is sometimes connected; it finally enters the skull by the foramen magnum, to join the *accessory* portion.

The spinal cord is supplied by several offsets from the vertebral arteries; near their termination in the basilar artery they give off two branches which unite under the name of the *anterior spinal artery;* this is continued to the bottom of the spinal canal by anastomoses from the vertebral arteries in the neck, and from the intercostal and lumbar arteries. The *posterior spinal artery* is also derived from the same source, and is continued down the posterior aspect of the spinal canal by anastomoses from the same branches that reinforce the anterior spinal artery.

The *veins* of the spinal cord are very tortuous, and form a plexus on its surface, emptying their contents into the vertebral, intercostal, lumbar, and sacral veins.

The spinal cord should be hardened in alcohol for examination, as, soon after death, it becomes softened, and unfit for dissection.

The SPINAL CORD extends from the medulla oblongata to the first or second lumbar vertebra. In shape it is a flattened cylinder, and it has two *enlargements*, the superior corresponding to the origin of the nerves for the upper extremity, and the inferior enlargement to that of those for the lower extremity. The cord has a fissure along its anterior surface, and another along its posterior surface; the anterior, called the *fissura longitudinalis anterior*, is the widest, and the posterior, called the *fissura longitudinalis posterior*, is the deepest; a *lateral fissure* also exists along the line of origin of the posterior roots of the spinal nerves, and another has been described as being found along the line of origin of the anterior roots.

A transverse section of the cord will show that each of its lateral halves is divided by the lateral fissure into two parts, that in front of the fissure being called the *antero-lateral column*, and that behind, the *posterior column*; it will also show that the two halves of the cord are united by a central portion which limits the depth of the longitudinal fissures, and is called the *commissure*.

A transverse section of the cord shows, also, that, like the brain, it is composed of white and gray substance, but the gray portion is surrounded by the white, instead of being external, as in the encephalon. The gray matter is arranged, in each half of the cord, in the form of a crescent, the horns of which point toward the roots of the nerves; the convexity looks toward the commissure, which also is chiefly made up of gray matter. The posterior horn of the crescent reaches to the fissure along the attachment of the posterior roots; the anterior horn does not reach to the anterior roots, nor does it form so sharp a point as the posterior.

The deep origin of the spinal nerves, like that of the cranial nerves, is uncertain.

DISSECTION VI.

SCAPULAR REGION.

The subject should be restored to its position with the face uppermost. All the muscles which attach the upper extremity to the thorax have, with a single exception, been examined, and if they have been divided it will be held only by the clavicle and serratus magnus muscle. The clavicle should be divided in the middle; the bundle of nerves and the artery cut through opposite the second rib, and tied in a bunch to the fragment of clavicle remaining. This exposes the serratus magnus muscle.

The SERRATUS MAGNUS MUSCLE covers a large portion of the thoracic parietes, and forms the inner wall of the axilla; it arises by nine muscular slips, arranged in a curved line, the convexity of which looks forward, from the anterior surface of eight upper ribs, two slips being attached to the second rib, and the lower of its slips indigitating with the external oblique muscle of the abdomen; it is inserted into the whole length of the posterior border of the scapula. The inferior thoracic and subscapular arteries ramify on the surface of this muscle, and the long thoracic nerve (p. 103) passes down from behind the axillary plexus, to be distributed to it.

The division of this muscle will complete the separation of the upper extremity from the thorax. In accomplishing this, it must be remembered that some important branches of the subclavian artery are distributed to the dorsum of the scapula, and these should be so divided as not to interfere with their further examination.

The SUBSCAPULARIS MUSCLE lies upon the inner surface of the scapula, covered in by a fibrous lamina which is but slightly adherent to it; this being removed, it will be found to arise from the inner surface of that bone, except at its inferior and superior angles; it is inserted by a broad flat tendon, which forms a part of the capsular ligament of the shoulder-joint, into the lesser tuberosity of the humerus. A band of fibres, two or three inches in length, is sometimes found extending from the scapula to the neck of the humerus, just below this muscle. The belly of this muscle is intersected longitudinally by aponeurotic laminæ, attached to the ridges of the scapula. The *subscapular nerve*, arising from the posterior part of the brachial plexus, will be seen entering this muscle immediately after its origin, penetrating the superior border near the commencement of its tendon; a small branch enters at the lower border, and a third, accompanying the subscapular artery, is eventually distributed to the latissimus dorsi muscle. The *subscapular artery* will also be noticed; passing along the lower border, it gives a branch to the deep surface of this muscle, and another to the dorsum of the scapula, and anastomoses with the posterior scapular branch of the subclavian at the inferior angle of the bone.

The TERES MAJOR MUSCLE lies below the sub-scapularis; it arises on the dorsum of the scapula from the flat surface constituting its inferior angle, and, leaving a triangular interspace between it and the lower border of the bone, is inserted by a broad tendon, conjoined with that of the latissimus dorsi, into the internal ridge of the bicipital groove of the humerus. A synovial bursa exists between the conjoined tendons.

The SUPRA-SPINATUS MUSCLE is situated on the dorsum of the scapula, above the spine of that bone; occupying the whole of the supra-spinous fossa, from the walls of which it arises; it is inserted into the upper facet of the greater tuberosity of the humerus, by a flattened tendon, which forms part of the capsular ligament of the shoulder-joint. Passing through the supra-scapular notch will be seen the *supra-scapular nerve*, a branch of the brachial plexus, which passes beneath this muscle to supply it, and then curves round the external border of the spine, to be distributed to the infra-spinatus muscle.

To trace this nerve, as well as to follow out the divided extremity of the supra-scapular artery, the acromion process should be sawed across at its base, and removed; the muscle is thus wholly displayed, and is to be divided near its tendon, and dissected out from the fossa, respecting all nervous and arterial branches which may be exposed.

The *supra-scapular artery* is a branch of the subclavian artery (p. 58), and passes over the ligament of the supra-scapular notch, to penetrate beneath the supra-spinatus muscle, which it supplies. A branch winds round the anterior border of the spine of the scapula, to inosculate with the dorsal branch of the subscapular, and with the branches of the posterior scapular distributed on the dorsum of the bone.

The INFRA-SPINATUS MUSCLE occupies the infra-spinous fossa, and is covered by a dense fascia, which, as well as the remains of the deltoid muscle, overlapping its anterior half, must be removed. It arises from the walls of the infra-spinous fossa, and from the fascia which covers it externally; it is inserted into the middle facet of the greater tuberosity of the humerus by a tendon which forms part of the capsular ligament of the shoulder-joint. Its tendon is at first concealed by the muscular fibres which overlap it, each half of the muscle being folded over the tendon, from which the fibres diverge in a bipenniform manner. Passing along its outer border is the *posterior scapular artery*, a branch of the subclavian; at the inferior angle of the scapula this artery inosculates with the subscapular, and by the offsets which it sends to the infra-spinatus muscle and fossa, unites with the terminal twigs of the supra-scapular.

The TERES MINOR MUSCLE lies between the infra-spinatus and the teres major muscles; it is closely connected with the former, and can only be separated from it artificially, so that the dissector is sometimes at a loss to define this muscle, though he may be perfectly aware of its locality. It arises from the inferior border of the scapula, and is inserted into the lower facet of the greater tuberosity of the humerus, its tendon forming part of the capsular ligament of the shoulder-joint. The *dorsal* branch of the subscapular artery curves around this muscle, just outside the scapular head of the triceps muscle, and passes to the infra-spinous fossa, beneath the infra-spinatus muscle, which it supplies, and where it inosculates with the terminal branches of the supra-scapular.

BACK OF THE ARM.

The TRICEPS EXTENSOR CUBITI MUSCLE makes up the whole bulk of the back of the arm; it has three points of origin, known as its long, middle, and short heads. The *long head* arises from the inferior border of the scapula, just below the glenoid cavity; the *middle head* arises from all the shaft of the humerus below its greater tuberosity, and from the external condyloid ridge and the intermuscular septum connected with it; the *short head* arises from the shaft of the humerus below the insertion of the teres major, and from the internal condyloid ridge and its intermuscular septum. These three heads conjoin, to be inserted by a broad aponeurotic tendon into the olecranon process of the ulna; between the tendon and the olecranon is a synovial bursa.

The long head of the triceps divides the triangular space left between the teres major and the subscapularis muscle into a smaller triangle on the outer side, and into a quadrangular space between it and the humerus on the inner side. Through the small triangular space passes the dorsalis scapulæ artery, and through the quadrangular space the posterior circumflex artery, and circumflex nerve. Beneath the belly of the triceps, and between it and the shaft of the humerus, pass the musculo-spiral nerve and the superior profunda artery.

The triceps muscle must be divided in the middle, and turned upward and downward to follow out the course of this artery and nerve.

DISSECTION VII.

FRONT OF THE FOREARM.

An incision should be made down the forearm to the wrist, and there joined by a short transverse one; the skin is to be removed in such a way as to permit the origin of the veins of the elbow to be seen, as well as the terminations of the cutaneous nerves (p. 107). The muscles of the forearm are surrounded by a firm aponeurosis, which not only invests them collectively, but penetrates between them individually. This aponeurosis is to be divided and removed; toward the condyles it will be found that the muscular fibres originate from it, and where they do, it will necessarily be left adherent to them. The muscles are best isolated from one another by commencing at the tendons, and tracing their separations upward to the elbow; the sheaths of the muscles and the cellular tissue lying in their interspaces must all be removed, and the tendons should be dissected as cleanly as possible; the beautiful appearance they present when properly dissected fully repays the labor spent upon them.

The tendon of the biceps muscle divides the muscles of the forearm into an external and an internal group, each group being collectively attached to the condyle of the humerus of its respective side by a *common tendon*, which also sends *septa* between the muscles; they are also divided into a deep and a superficial layer. In separating these muscles, the arteries and nerves will necessarily come into view. It will be seen that the brachial artery dips downward between the muscles, and divides into two branches, the radial and ulnar; these pass down the arm between the muscles to the wrist, where they become comparatively superficial, and then enter the hand. The median nerve lies upon the inner side of the artery, at the hollow of the elbow, and afterward passes down the middle line of the limb in its course to the hand; the ulnar nerve will be found on the outside of the ulnar artery in its lower two-thirds; the radial nerve also accompanies the radial artery till within three inches of the wrist, where it becomes cutaneous, and divides into two branches, distributed respectively to the back of the thumb and dorsum of the hand.

The PRONATOR RADII TERES is the first muscle next the tendon of the biceps on its inner side; it arises by two heads, one from the inner condyle, and from the common tendon above mentioned, the other, deeper, and not to be seen in the present stage of the dissection, from the coronoid process of the ulna; the condyloid attachment sometimes receives additional fibres from the intermuscular septum above the condyle; this peculiarity is usually associated with the existence of a supra-condyloid process (p. 106). The muscle passes obliquely outward, to terminate in a flat tendon which winds round the radius, and is inserted into a rough surface on its outer side; this insertion cannot be seen until the superficial muscles are removed. The *median nerve* passes between its two heads.

The FLEXOR CARPI RADIALIS, arising next the pronator radii teres, from the inner condyle and the common tendon, becomes tendinous near its middle, and passing through a distinct sheath, outside the arch of the annular ligament, is inserted into the base of the metacarpal bone of the index finger. The insertion cannot be seen till the hand is dissected. The *radial artery* passes along the outer border of the lower part of its tendon.

The PALMARIS LONGUS MUSCLE lies on the inner side of the flexor carpi radialis; it has a small belly and a long tendon, and is very often wanting; it arises from the internal condyle and common tendon, and continuing down the centre of the forearm and over the annular ligament, is inserted into the palmar fascia, with which it is continuous.

The FLEXOR CARPI ULNARIS passes along the ulnar border of the forearm, arising from the internal condyle and common tendon, and also from the inner edge of the olecranon, by a strong but thin aponeurosis, underneath which pass the ulnar nerve, and the recurrent branch of the ulnar artery; its tendon receives short muscular fibres nearly down to the point of its insertion, which is into the pisiform bone and base of the metacarpal bone of the little finger. At its upper part this muscle overlaps the ulnar artery and nerve; below the middle they are upon the inner side of its tendon.

The flexor carpi radialis and palmaris longus muscles must be divided in their middle and the two ends reflected, as, at their upper part, they cover up the muscle next to be dissected. The numerous branches of the ulnar artery at the elbow should be carefully preserved.

The FLEXOR SUBLIMIS DIGITORUM forms a large part of the muscular mass arising from the inner condyle; it also arises from the coronoid process of the ulna and the oblique line of the radius; inferiorly it divides into four tendons which pass beneath the annular ligament to be inserted into the bases of the second phalanges of the fingers, as will be seen in the dissection of the hand. Beneath the annular ligament the tendons are provided with a synovial membrane. A muscular slip often connects this muscle with the flexor profundus, or the flexor longus pollicis.

The upper part of the flexor sublimis, where it arises from the radius, is covered by the pronator radii teres, the tendinous insertion of which, winding round the radius, as well as its coronoid head, can now be seen. The tendon of the *biceps* may also be followed between the muscles to its insertion into the tubercle of the radius. The *brachialis anticus* will likewise be exposed so that its insertion into the coronoid process of the ulna can be examined. None of these insertions could be seen when the bellies of these muscles were dissected.

The SUPINATOR LONGUS MUSCLE gives the rounded outline characteristic of the outer side of the forearm; it arises from the humerus on its outer side, nearly as high as the insertion of the deltoid muscle, and from the external condyloid ridge; it passes down the radial side of the forearm and is inserted by a flattened tendon into the external border of the radius, just above the base of its styloid process.

The RADIAL ARTERY, with its venæ comites, lies upon the inner side of the last-named muscle at its upper part, and between it and the pronator radii teres; lower down it lies between the supinator longus and flexor carpi radialis; at its upper part it gives off the *radial recurrent branch*, which, turning backward beneath the belly of the supinator longus, sends off numerous muscular twigs and inosculates with the superior profunda branch of the brachial; in its course to the wrist it gives off many *muscular offsets*, and, at the wrist, an *anterior* and *posterior carpal branch*, which pass transversely across in front and behind, to anastomose with similar branches from the ulnar artery. The *superficialis volæ* branch arises from an uncertain point near the wrist, and passes on to the ball of the thumb, there to lose itself in the muscles or to join with the superficial palmar arch; when given off high up, it occasionally furnishes one or two digital branches. Having reached the wrist, the radial artery winds round the base of the metacarpal bone of the thumb, beneath its extensor tendons, to enter the palm of the hand between the two heads of the first dorsal interosseous muscle; occasionally it curves round the radius higher up than this, and sometimes it passes directly over the annular ligament into the palm; its whole course may also be superficial, owing to its high division from the brachial (p. 106). The radial artery oftentimes presents, especially in old and fat subjects, a series of flexuosities attended by dilatation; this condition is usually accompanied by a deposit of calcareous matter in the arterial walls, varying in quantity; the same thing may be noticed in other arteries, and in these cases, ossific, or atheromatous deposits, will be found to a considerable extent in the upper part of the aorta.

The RADIAL NERVE is the larger of the two branches into which the musculo-spiral nerve divides in front of the external condyle; it accompanies the radial artery upon its outer side beneath the supinator longus muscle; near the wrist it passes under the tendon of the supinator, becomes cutaneous, and divides into two branches, one for the back of the thumb, and the other for the back of the hand.

The MEDIAN NERVE passes between the two heads of the pronator radii teres, and beneath the flexor sublimis digitorum, where it gives off the anterior interosseous and muscular branches; near the wrist it becomes superficial, appearing along the outer border of the tendons of the latter muscle; it here gives off a *superficial palmar* branch, which passes over the annular ligament to the muscles and integument of the ball of the thumb, while the main part of the nerve continues beneath the ligament to the fingers.

When the brachial artery passes behind the supra-condyloid process (p. 106), the median nerve always follows it, but the nerve may curve around the process without the artery; if the process exists the nerve invariably deviates from its course, the artery generally, but not always. The median nerve is sometimes accompanied by an artery of considerable size, called the *median artery*, given off by the anterior interosseous or ulnar; it accompanies the nerve to the hand, where it joins one of the palmar arches, or one of the digital branches.

The ULNAR ARTERY, at its origin from the brachial, lies upon the brachialis anticus muscle; it then dips beneath the flexor sublimis digitorum, crosses obliquely to the inside of the arm, and, at its middle third, becomes more superficial, lying between the tendons of the flexor sublimis and flexor carpi ulnaris, to which it gives several muscular twigs; it crosses the annular ligament under a strong fascia thrown over it from the pisiform bone, and there forms the superficial palmar arch, covered in by the palmar fascia. Just beyond its origin it gives off the *anterior ulnar recurrent* branch, which, passing backward between the brachialis anticus and pronator radii teres muscles, breaks up into muscular branches, and inosculates with the inferior profunda and anastomotica magna of the brachial. The *posterior ulnar recurrent* branch sometimes originates by a common trunk with the preceding, and sometimes is given off a little lower down; it passes beneath the superficial muscles of the inside of the forearm, and emerging beneath the tendon of the flexor carpi ulnaris at the side of the ulnar nerve, anastomoses with the inferior profunda and anastomotic arteries. The *common interosseous* artery is given off just below these branches, and its divisions will be hereafter described. The ulnar artery gives off only muscular branches until it reaches the wrist, where it furnishes an *anterior carpal* branch to the front, and a *posterior carpal* branch to the back of the wrist, both of which pass transversely across to anastomose with similar branches from the radial. In cases of high division, the ulnar artery is usually superficial in the forearm.

The ulnar artery is accompanied by two venæ comites.

The ULNAR NERVE, after passing under the origin of the flexor carpi ulnaris, continues beneath that muscle to about the middle of the forearm, where it joins the artery, and descends along its outer side to the wrist: it here gives off a branch which supplies the back of the hand, and then, with the artery, passes over the annular ligament to the palm of the hand.

The flexor sublimis, flexor carpi ulnaris, and supinator longus muscles, are now to be divided across their tendons, and their muscular bellies reflected. The pronator radii teres may be drawn to one side by hooks. The arteries and nerves should remain undivided.

The FLEXOR PROFUNDUS DIGITORUM lies upon the ulna, and arises from the upper two-thirds of that bone, and from the interosseous membrane; it divides into four tendons, which are not, however, separable above the annular ligament, beneath which they pass, and are inserted into the bases of the last phalanges, having perforated the tendons of the flexor sublimis.

The FLEXOR LONGUS POLLICIS lies beneath the supinator longus muscle, and upon the radius; it arises from the upper two-thirds of that bone, from the coronoid process of the ulna, and from the interosseous membrane; its tendon passes beneath the annular ligament, and is inserted into the last phalanx of the thumb.

The COMMON INTEROSSEOUS ARTERY, after arising from the ulnar, quickly divides into two branches, anterior and posterior. The *anterior branch* passes down the arm, between or in the deep flexor muscles, in close relation to the interosseous membrane; beneath the pronator quadratus muscle the artery passes through this membrane to anastomose with the posterior carpal branches of the ulnar and radial. The *posterior branch* passes through the interosseous membrane at its upper part, and is distributed to the posterior aspect of the arm.

The *anterior interosseous nerve*, a branch of the median, accompanies the anterior interosseous artery, and terminates in the pronator quadratus muscle.

The PRONATOR QUADRATUS MUSCLE is a flat quadrilateral muscle, stretched transversely across the lower part of the bones of the forearm; it arises from the anterior surface and border of the ulna, and is inserted into the anterior surface of the radius; the insertion is usually a little narrower than its origin.

DISSECTION VIII.

BACK OF THE FOREARM AND HAND.

The skin is to be removed from the back of the arm and hand. The muscles are covered in by a dense fascia continuous with that of the front of the arm; this may be removed at its lower part, but it is adherent to the bellies of the muscles above; the muscles should be separated by tracing them upward from their tendons. The back of the arm is more difficult to dissect neatly than the front.

Before commencing the dissection of the muscles, the *dorsal branch of the radial nerve* should be followed out. It becomes cutaneous at the lower third of the radius, and divides into two branches; one of which is distributed to the radial border and ball of the thumb; the other divides into *dorsal digital branches* and supplies the remaining side of the thumb, both sides of the next two fingers, and half the ring finger. The *dorsal branch of the ulnar nerve*, appearing near the styloid process of the ulna, supplies both sides of the little finger, and the contiguous side of the ring finger upon their dorsal surfaces.

The muscles of the back of the arm are divided into two layers; the separation is not well defined, and the muscles are not so voluminous as those of the front of the arm.

The EXTENSOR CARPI RADIALIS LONGIOR lies upon the radial side of the arm, just below the supinator longus, by which it is partly covered; it arises from the external condyloid ridge of the humerus, and its tendon passes through a well-marked groove in the head of the radius, which is covered by the posterior annular ligament, to be inserted into the base of the metacarpal bone of the index finger. The radial nerve lies along the outer border of its tendon.

The EXTENSOR CARPI RADIALIS BREVIOR immediately succeeds the preceding muscle, and is partly covered by it; it arises from the outer condyle of the humerus, and the tendon common to the extensor muscles; it forms a tendon closely united with that of the extensor carpi radialis longior, and passes through the same groove in the radius, beneath the annular ligament; after which it diverges from it, and is inserted into the base of the metacarpal bone of the middle finger. The tendons of both these muscles pass beneath the extensor tendons of the thumb.

The EXTENSOR COMMUNIS DIGITORUM occupies the central portion of the posterior region of the forearm; it arises from the external condyle by the tendon common to the extensor muscles, and from the intermuscular septa between it and the contiguous muscles; at the lower part of the arm it divides into three tendons, which pass through the annular ligament in a compartment with the extensor indicis; escaping from the ligament, the most internal tendon divides into two, and the four tendons pass along the dorsum of the hand, forming a flattened sheath for the back of each finger. Opposite the first phalangeal articulation this expanded tendon divides into three slips; the central one is inserted into the base of the second phalanx, the two lateral continue onward and are inserted into the dorsal surface of the last phalanx. Oblique tendinous bands connect the tendons with each other on the back of the hand, and upon the fingers they are reinforced by tendinous slips from the lumbricales and interossei muscles.

The EXTENSOR MINIMI DIGITI is generally a part of the extensor communis; occasionally it is separable from it. Its origin is the same, and it passes through a separate ring of the annular ligament; its tendon, which is split into two directly afterward, terminates in an expansion on the back of the little finger.

The EXTENSOR CARPI ULNARIS arises from the external condyle, and the common tendon of the extensors, and from the upper part of the ulna; its tendon passes through a separate sheath of the annular ligament, just over the carpal end of the ulna, to be inserted into the base of the metacarpal bone of the little finger.

The ANCONEUS MUSCLE is a small triangular muscle placed upon the posterior part of the elbow-joint, which it partly covers, and is sometimes considered as a part of the triceps extensor cubiti; it arises from the outer condyle by a distinct tendon posterior to the common tendon of the extensor muscles, and is inserted into the radial side of the olecranon and the adjacent surface of the ulna.

The extensor muscles of the arm, which have been described, are now to be divided in the middle, and their two ends reflected; however careful he may have been, the student must expect the deep layer of muscles to present a ragged appearance.

The *posterior interosseous artery* perforates the interosseous membrane at its upper part, and appears between the supinator brevis and the extensor ossis metacarpi muscles; it descends between the deep and superficial layers of muscles, supplying them with muscular branches, and anastomoses with the posterior carpal arteries of the radial and ulnar, and with the terminal twigs of the anterior interosseous; at its upper part it gives off a *recurrent* branch, which passes beneath the anconeus muscle to supply the elbow-joint, and anastomose with a branch of the superior profunda of the brachial.

The *posterior interosseous nerve* is given off, in front of the outer condyle, from the musculo-spiral nerve; it passes through the fibres of the supinator brevis to descend the back of the arm between the two layers of muscles as far as the middle of the forearm, where it sinks beneath the extensor secundi internodii pollicis, and is distributed to the back of the carpus; sometimes it has a gangliform swelling at its termination.

The SUPINATOR BREVIS MUSCLE is a small muscle at the upper part of the arm, the fibres of which pass obliquely round the upper third of the radius; it arises from the external lateral and the orbicular ligaments, and from about two inches of the upper part of the ulna, and is inserted into the oblique line on the upper part of the radius, except at its inner part.

The EXTENSOR OSSIS METACARPI POLLICIS lies next below the supinator brevis, and is sometimes united with it; it arises from the posterior surface of the radius, from the ulna, and from the intervening interosseous membrane; it forms a large belly, the tendon of which, passing through the outer compartment of the posterior annular ligament, and in a groove of the radius common to it and the extensor primi internodii pollicis, is inserted into the base of the metacarpal bone of the thumb.

The EXTENSOR PRIMI INTERNODII POLLICIS is the smallest muscle of the deep layer, and its tendon is closely connected with that of the preceding muscle; it arises from the radius and interosseous membrane just below the origin of that muscle, and passing through the same compartment in the annular ligament, is inserted into the base of the first phalanx of the thumb.

The EXTENSOR SECUNDI INTERNODII POLLICIS lies next below, and is partly covered by the preceding muscle; it arises from the posterior surface of the ulna, and from the interosseous membrane, and descends obliquely to the thumb; its tendon, crossing the radial artery and the extensor muscles of the wrist, passes through a special sheath in the annular ligament, and in a groove in the radius, to be inserted into the base of the second phalanx of the thumb. The little triangular interval left between the tendon of this muscle and the parallel tendons of the two preceding muscles, has been called the "anatomist's snuff-box;" forced abduction of the thumb will reveal the depression to which this name is given.

The EXTENSOR INDICIS is the lowest muscle toward the wrist; it arises from the shaft of the ulna, usually below the middle; its tendon passes through the annular ligament with the common extensor of the fingers, and uniting with the tendon of that muscle going to the forefinger, is inserted with it into the second and third phalanges.

The POSTERIOR ANNULAR LIGAMENT is formed from the deep fascia of the arm, and, if divided over the tendons passing through it, will be found to have six separate canals, lined with synovial membrane; with the exception of that for the extensor minimi digiti, which lies in the interval between the two bones, each of these canals has a corresponding groove in the radius or ulna; between each of the canals, the ligament is firmly attached to the bone beneath.

The *radial artery* will be found passing round the lower end of the radius, beneath the extensor tendons of the thumb, to the space between the thumb and forefinger, where it passes between the two heads of the first dorsal interosseous muscle, and penetrates the palm of the hand. Before disappearing, it gives off the *dorsal carpal branch*, which crosses transversely beneath the extensor tendons, to join a similar branch from the ulnar artery. From the arch thus formed, *dorsal interosseous arteries* are given off to the third and fourth interosseous spaces; the *metacarpal*, or *first dorsal interosseous*, larger than the others, passes upward, in the space between the first and second metacarpal bones, anastomosing, as do the others, with the perforating branch of the deep palmar arch; at the cleft of the index and middle fingers it ends by joining with the digital branch of the superficial palmar arch. The radial artery also gives off two small dorsal branches to the thumb, and a dorsal branch to the index finger.

In addition to its posterior carpal branch, the ulnar

artery sends a *metacarpal branch* along the metacarpal bone of the little finger.

The DORSAL INTEROSSEOUS MUSCLES are four in number, occupying the spaces between the metacarpal bones, and arising by a double head from the lateral surfaces of the two bones between which they lie. The first is larger than the others, and is called the *abductor indicis;* it is inserted into the radial side of the first phalanx and the extensor tendon of the index finger; the radial artery passes between its heads; the second terminates in the first phalanx and extensor tendon of the middle finger on the radial side; the third is also inserted into the first phalanx and extensor tendon of the middle finger, but upon its ulnar side; the fourth is inserted into the first phalanx and extensor tendon of the ring finger on its ulnar side.

DISSECTION IX.

PALM OF THE HAND.

To dissect the hand, the fingers and thumb should be separated widely from each other and fastened to the table by pins driven through the skin at the tip of each finger. An incision made down the middle of the palm, and another transversely across the roots of the fingers, from which others can be carried down the length of each, will permit the removal of the thick skin and fat covering and penetrating the perforations of the dense fascia radiating from the annular ligament to the fingers. Upon the ulnar side the student must respect the palmaris brevis muscle, composed of a few bundles of transverse fibres lying in the fat just beneath the integument, and he must be careful not to destroy the transverse ligament, which stretches between the commissures of the fingers.

The PALMARIS BREVIS MUSCLE consists of a small though variable number of transverse fibres, arising from the palmar fascia and annular ligament; it is inserted into the integument of the ulnar border of the hand.

The *ulnar nerve*, dividing on the annular ligament, sends a *superficial branch* to the palmaris brevis and gives off *two digital nerves* which supply both sides of the little finger and the contiguous side of the ring finger. A deep branch penetrates between the abductor and flexor minimi digiti muscles, then passes under the flexor tendons, accompanying the deep palmar arch, and is distributed to the interossei and lumbricales muscles, and to the muscles of the thumb.

The *median nerve* sends an offset across the annular ligament which passes down the middle of the palm and unites with one of the superficial branches of the ulnar nerve.

The PALMAR FASCIA, continuous with the anterior border of the annular ligament, spreads out anteriorly, covering in the tendons and vessels, and opposite each finger divides into slips attached to the sides of the first phalanges; between these, emerge the tendons, nerves, and vessels of the fingers; strong transverse fibres attached to the phalanges on each side, called the *transverse ligament*, form a framework over which is stretched the skin constituting the commissures of the fingers. The palmar fascia is held down by fibres attached along the metacarpal bones.

The palmar fascia is to be removed with the scissors, preserving the nerves and arteries beneath it; the examination of the muscles of the thumb is then to be proceeded with.

The ABDUCTOR POLLICIS is the most superficial muscle of the thumb; it is flat and narrow, and arises from the annular ligament and the os trapezium; it is inserted into the base of the first phalanx of the thumb; it is oftentimes connected at its origin with a slip from the tendon of the extensor ossis metacarpi pollicis.

The FLEXOR OSSIS METACARPI POLLICIS, or OPPONENS POLLICIS MUSCLE, lies beneath the preceding, its fibres projecting on both sides; it arises from the annular ligament and os trapezium, and is inserted into the whole length of the metacarpal bone of the thumb.

The superficialis volæ branch of the radial artery crosses these muscles at their origin, and either terminates in their fibres, or continues on to inosculate with the terminal part of the superficial arch of the ulnar artery. It is not always present.

The FLEXOR BREVIS POLLICIS is the largest of the muscles of the thumb; it has two points of origin, one from the annular ligament and trapezium, the other from the os trapezoides and os magnum; between these two portions passes the tendon of the long flexor of the thumb, which it will be remembered is a brachial muscle (p. 145); they then unite into one mass, to be inserted by two heads into the sides of the base of the first phalanx of the thumb, the inner being united with the adductor pollicis, and the outer with the abductor pollicis; a sesamoid bone is connected with each at its insertion. A branch from the *median nerve* supplies the outer part of the flexor and the abductor and opponens pollicis muscles.

The examination of the adductor pollicis muscle is necessarily deferred to a later period of the dissection.

The palmaris brevis being removed, the most superficial of the remaining muscles of the little finger is the ABDUCTOR MINIMI DIGITI; it arises from the pisiform bone and the tendon of the flexor carpi ulnaris, and is inserted into the ulnar side of the base of the first phalanx of the little finger. This muscle may arise from the fascia of the forearm with a length of four inches. A branch from the ulnar artery, called the *communicating*, and a *deep branch* of the ulnar nerve, pass between this muscle and the one beneath it.

The FLEXOR BREVIS MINIMI DIGITI lies beneath the preceding, of which it appears to be a part; it arises from the unciform bone and from the annular ligament, and is inserted into the base of the first phalanx of the little finger. This muscle is sometimes wanting.

The FLEXOR OSSIS METACARPI, ADDUCTOR, or OPPONENS MINIMI DIGITI is partly overlaid by the preceding muscles; it arises from the annular ligament and from the process of the unciform bone, and is inserted into the whole length of the metacarpal bone of the little finger.

The continuation of the ulnar artery in the palm of the hand is called the *superficial palmar arch;* it is covered in by the palmar fascia, and lies across the flexor tendons of the fingers and the digital branches of the median nerve; it gives off a *deep communicating* branch, which passes between the abductor and flexor minimi digiti, to inosculate, as will be seen hereafter, with the deep arch of the radial artery; it supplies both sides of the three inner fingers, and one side of the index finger, with *digital* branches; these arise opposite the interosseous spaces by single trunks, which bifurcate to supply the contiguous sides of two fingers, those of the same finger uniting at its extremity; the branch to the outer side of the little finger arises singly. Near the roots of the finger the digital arteries receive *communicating* branches from the deep arch; but the artery to the little finger gets its communicating branch from the deep arch about the middle of the hand. The terminal part of the superficial artery unites with the superficialis volæ, and with the branch of the radial artery which supplies the radial side of the forefinger.

The *median nerve*, after emerging from beneath the annular ligament, divides into two trunks; the external of these, besides giving off a muscular branch to the muscles of the thumb (p. 143), divides into three digital branches, which go to the two sides of the thumb and the radial side of the forefinger; the internal trunk divides into two branches which bifurcate to supply the contiguous sides of the fore and middle, and middle and ring fingers. The *digital* nerves pass down the sides of the fingers, superficially to the arteries, and terminate by filaments in the pulp at their extremities.

The ANTERIOR ANNULAR LIGAMENT is a firm ligamentous band, beneath which pass the tendons of the flexor muscles; it is attached to the trapezium and scaphoid bones on one side, and to the unciform and the pisiform bones on the other. The canal formed by this ligament is lubricated by a synovial membrane, which surrounds each tendon separately, and is prolonged both above and below the ligament, sometimes even as far as the fingers, where it communicates with the synovial membrane of the sheaths of some of the tendons; this communication has been demonstrated to be nearly constant in the tendons of the thumb and little finger; with the others it is only of occasional occurrence. This continuity in their synovial membranes explains the terrible consequences which sometimes follow phlegmonous inflammation of the fingers. The synovial sac of the wrist and fingers may be demonstrated by insufflation.

The annular ligament is now to be divided ; the median and ulnar nerves are also to be cut through, and turned over toward the fingers.

The *tendons of the flexor sublimis digitorum* muscle are superficial to those of the deep; these enter the sheaths of the fingers, and are inserted by two processes into the margins of the second phalanges at about their middle, being split opposite the first phalanges for the passage of the deep flexor tendons. The *sheath* of the tendons consists of transverse tendinous fibres attached to the sides of the first and second phalanges; in front of the articulations the sheaths are either wanting or imperfectly developed; they are lined by a synovial membrane.

The tendons of the flexor sublimis are to be divided and reflected toward the fingers.

The LUMBRICALES MUSCLES are small and delicate muscular slips connected with the deep flexor tendons; they are usually four in number, sometimes only three, and occasionally as many as six; they arise from the radial side of the tendon, near the annular ligament, and are partly concealed by the tendons of the flexor sublimis digitorum; they are inserted into the tendinous expansions of the extensor muscles covering the backs of the fingers.

The *tendons of the flexor profundus digitorum* muscle, the division of which takes place in the palm, enter the sheaths of the tendons of the flexor sublimis, pass through the split in these, and continue onward to be inserted into the bases of the last phalanges. They are attached to the phalanges and the capsules of the finger-joints along the median line, by membranous folds, given off from their posterior surface, and which contain elastic tissue; these are called the *ligamenta brevia*, or *vincula subflava*, and are supposed to hold the tendons down when the fingers are bent.

The tendon of the flexor longus pollicis passes through the annular ligament, externally to those of the flexor profundus, and turns outward between the heads of the flexor brevis pollicis to be inserted into the last phalanx of the thumb.

The deep flexor tendons are to be divided and reflected toward the fingers.

The ADDUCTOR POLLICIS MUSCLE, obscured by the flexor tendons at an earlier period of the dissection, is now seen arising from the anterior two thirds of the metacarpal bone of the middle finger, on its anterior aspect; its fibres converge to form a small tendon inserted into the inner side of the first phalanx of the thumb.

The *radial artery* enters the hand at the first interosseous space, between the two heads of the abductor indicis muscle. It furnishes a branch to the thumb, called the *princeps pollicis;* this divides into two branches, which pass along its sides and inosculate in its pulp; another branch, either from the radial or from the princeps pollicis, supplies the radial side of the forefinger, and is known as the *radialis indicis;* it unites at the end of the finger with the digital branch furnished to the opposite side by the ulnar artery. At the anterior border of the adductor pollicis, the radialis indicis communicates with the superficial palmar arch. The continuation of the radial artery is called the *deep palmar arch;* it extends across the interosseous muscles and the metacarpal bones to the little finger, where it anastomoses with the deep communicating branch of the ulnar artery. The arch sends off *recurrent branches*, which pass backward to the carpus, and *perforating arteries*, which penetrate the interosseous muscles to join the interosseous arteries on the back of the hand. Usually there are, also, three *palmar interosseous* arteries which pass forward and unite with the digital arteries of the superficial arch at the cleft of the fingers.

Although the distribution of the arteries of the hand, which has been described, is the one considered normal, it is to be remembered that "it is not possible to know, before a hand is opened, in what manner the arteries are distributed."[1] Sometimes the radial furnishes the superficial arch, and sometimes there is no arch at all; the arrangement is extremely irregular, and the dissector must be prepared to find a very different arrangement from that described.

The *deep branch of the ulnar nerve* accompanies the deep palmar arch across the metacarpal bones to the muscles of the thumb, and terminates in branches to the adductor pollicis, the inner head of the short flexor, and the abductor indicis.

The PALMAR INTEROSSEOUS MUSCLES are three in number, and are placed upon, rather than between, the metacarpal bones. The *first* arises from the second metacarpal bone, and is inserted into the ulnar side of the first phalanx and extensor tendon of the forefinger; the *second* arises from the fourth metacarpal bone, and is inserted into the radial side of the first phalanx and extensor tendon of the ring-finger; the *third* arises from the metacarpal bone of the little finger, and is inserted into the radial side of the first phalanx and extensor tendon of the little finger.

The interosseous and lumbricales muscles being supposed to convey to the hand that peculiar dexterity and delicacy of use which is found only in certain professions or individuals, the dissecting-room is not the place to find them in their fullest degree of development.

---

[1] Horner.

DISSECTION X.

LIGAMENTS OF THE RIBS, SPINE, AND UPPER EXTREMITY.

To dissect the ligaments, all the fleshy part of muscles surrounding the joints must be removed; the tendinous insertions are to be left, as in many instances they enter largely into the formation of the articulations; the cellular and all other extraneous tissues are to be cleared away, so that nothing shall be left to obscure the pearly aspect which these parts present when properly dissected. Ligamentous preparations are capable of very brilliant display, and, though requiring much patience, are extremely interesting to make.

A piece of the spinal column, with three or four ribs attached, furnishes the means of examining the COSTO-VERTEBRAL and VERTEBRAL ARTICULATIONS.

The ribs are attached to the vertebræ by two groups of ligaments; one extending from the head of the rib to the bodies of the vertebræ, and the other from the tubercle of the rib to the transverse processes.

The head of the rib, except that of the first, eleventh, and twelfth ribs, is received in a hollow on the sides of the bodies of two contiguous vertebræ, and is held in place by the *stellate ligament*, which passes from it to the vertebra in a radiated manner, and in those ribs connected with two vertebræ consists of three distinct portions, one for the superior, and one for the inferior vertebra, and a central portion to the inter-vertebral fibro-cartilage.

The inter-articular ligament can only be seen by a vertical section, which shall include the bodies of the contiguous vertebræ, and the neck of the rib.

The *inter-articular ligament* is a short, thin band attached on the one side to the ridge which separates the head of the rib into two articulating surfaces, and on the other, to the inter-vertebral fibro-cartilage; it divides the joint into two cavities, each furnished with a synovial sac; the first, eleventh, and twelfth ribs have no inter-articular ligament, and consequently but one synovial sac.

The *costo-transverse ligament* extends from the neck and tubercle of the rib to the transverse process of the vertebra; the *anterior* ascends from the upper border of the neck of the rib to the lower border of the transverse process of the upper of the two vertebræ with which it is connected; it is necessarily wanting in the first rib.

Between this ligament and the vertebra, emerges the posterior branches of the intercostal artery and nerve. The *posterior* costo-transverse ligament extends from the tubercle of the rib to the tip of the transverse process.

The middle costo-transverse ligament can only be seen by a horizontal section, made through the rib and transverse process, across the vertebra.

The *middle*, or *interosseous transverse ligament*, is a very strong band, passing directly between the posterior surface of the neck of the rib, and the surface of the transverse process against which the rib rests. A synovial sac is found between the tubercle of the rib and the transverse process, except in the lower two ribs, where the tubercles and transverse processes do not touch.

The several vertebræ of the spinal column are united by ligaments between their bodies and processes; these correspond throughout the column, except between the first two vertebræ and the head; these latter are described at p. 74.

The *anterior common ligament* is a broad glistening band which reaches the whole length of the vertebral column; it rests upon the front of the bodies of the vertebræ; its fibres run longitudinally, and are widest opposite the lumbar vertebræ.

The *posterior common ligament* lies upon the posterior aspect of the bodies of the vertebræ, within the vertebral canal, reaching from the sacrum to the occipital bone; its fibres run longitudinally, and are in contact with the dura mater; it is widest opposite the cervical vertebræ, and as it expands opposite each inter-vertebral disk, it has a scolloped outline along its borders.

The *inter-vertebral substance* is displayed by separating two vertebræ; it lies between the contiguous surfaces of their bodies, in the form of a circular disk, acting as a "buffer" against the shocks to which the vertebræ would otherwise be subject, and consists of a firm, outer, fibrous portion, the layers of which are concentrically arranged, and of a soft and elastic central portion, which bulges when two vertebræ are cut apart, or sawn through longitudinally. By separating the inter-vertebral substance from the bone, it will be found that the vertebra has a cartilaginous covering between it and the disk.

The processes of the vertebræ have special uniting ligaments.

The *ligamenta subflava* are elastic layers placed between the arches of the vertebræ, stretching from the lower border of one to the upper border of the next; they are longest in the cervical region.

The *supra-spinous ligament* extends along the tips of the spinous processes, and is best developed in the lumbar region.

The *inter-spinous ligaments* lie between the upper and lower borders of the spinous processes, and extend from the root to the tip of the process; they are best marked in the lumbar region.

The *inter-transverse ligaments* extend between the transverse processes, as thin bands in the lumbar, and as round bundles in the dorsal vertebræ; they are wanting in the cervical region.

A *capsular ligament*, not very well marked, surrounds the articular processes, and incloses a synovial membrane.

The clavicle is united to the scapula by an articulation with the acromion, and by a ligament between the clavicle and the coracoid process.

The ACROMIO-CLAVICULAR ARTICULATION is maintained by scattered fibres, which make a kind of capsule for the joint; an inter-articular fibro-cartilage, often indistinct, exists between the two bones, and a synovial membrane lines the interior of the articulations.

The *coraco-clavicular ligament* is a thick fasciculus, reaching from the base of the coracoid process to the under surface of the clavicle; when seen from the front, its fibres present a quadrilateral shape, and it is hence called the *trapezoid ligament;* when seen from behind, it is triangular in shape, and is then called the *conoid ligament.*

The scapula has two ligaments unconnected with any other bone; the coraco-acromial, and the transverse.

The *coraco-acromial ligament* is triangular in shape, and extends transversely between the coracoid process and the acromion, its apex being attached to the end of the acromion, and its wider base into the whole length of the coracoid process.

The *transverse ligament* converts the notch in the upper border of the scapula into a foramen.

A *capsular ligament* incloses the SHOULDER JOINT. It is attached above, to the neck of the scapula, and below, to the humerus, close to its articular surface; internally it is lined with a synovial membrane; it is strengthened by the tendons of the muscles of the scapula, and by a broad band, called the *coraco-humeral ligament*, which extends from the base of the coracoid process to the greater tuberosity of the humerus. The long tendon of the biceps muscle penetrates the capsule between the tuberosities of the humerus, and is attached to the upper part of the glenoid fossa of the scapula; this tendon is surrounded by a prolongation of the capsular synovial membrane.

The *glenoid ligament* consists of a fibrous band, continuous with the tendon of the biceps, from which it seems to be formed, and which surrounds and deepens the glenoid fossa of the scapula.

The bones of the ELBOW-JOINT are kept in place by the following ligaments:—

The *external lateral ligament* consists of a round fasciculus attached to the external condyle above, and the orbicular ligament of the radius below.

The *internal lateral ligament*, triangular in shape, is attached above, by its apex, to the internal condyle, and below, by its base, to the margin of the sigmoid notch of the ulna, from the coronoid process to the olecranon.

When the supra-condyloid process is present, a ligament extends from its tip to the internal condyle, thus completing the analogy it is supposed to have with the foramen in the lower part of the humerus, through which the brachial artery passes in certain classes of animals.

The *anterior ligament* extends from the front of the humerus to the coronoid process and orbicular ligament; its fibres are very thin and pass in various directions.

The *posterior ligament* is attached superiorly to the humerus, above the fossa for the olecranon, and inferiorly to the edges of the olecranon process.

The radius is held to the ulna by the *orbicular ligament*, a broad band which surrounds its head, and is inserted by its two extremities at either end of the lesser sigmoid notch of the ulna.

The *oblique ligament* is a slender band, sometimes wanting, which extends from the front of the coronoid process to the radius below its tubercle.

The *tendon of insertion of the biceps muscle* may now be examined better than could be done previously; a bursa is found between it and the bone; near its attachment the tendon twists, its anterior surface becoming external, and *vice versâ*.

The *interosseous membrane* is a thin fibrous layer, attached to the contiguous margins of the radius and ulna, separating the muscles of the front and back part of the forearm; its fibres are directed obliquely downward toward the ulna; superiorly the membrane is wanting.

The WRIST-JOINT is maintained by four ligaments:—

The *external lateral ligament* is a short strong band between the styloid process of the radius and the upper part of the scaphoid bone.

The *internal lateral ligament* is smaller but longer than the external; it extends between the styloid process of the ulna and the upper part of the cuneiform bone.

The *anterior ligament* is a membranous layer reaching from the end of the radius to the anterior surface of the first row of carpal bones.

The *posterior ligament* is also a membranous layer, and extends from the lower end of the radius to the posterior aspect of the first row of carpal bones.

The radius and ulna are held together at their lower articulation by a *triangular fibro-cartilage*, placed between them; this is attached by its apex to the inner surface of the styloid process of the ulna, and by its base to the edge of the lesser articulating surface of the radius; a few scattered fibres loosely surround this joint by way of a capsular ligament.

The carpal bones are united into two rows by *dorsal, palmar*, and *interosseous bands*, and the two rows are similarly united with each other; they are all supplied by one synovial membrane, except the *pisiform bone*, which has a *capsule* and *synovial* membrane distinct from the others; it has also *two special ligaments*, one to the process of the unciform bone and the other to the base of the fifth metacarpal bone.

The CARPO-METACARPAL ARTICULATIONS are maintained by *dorsal* and *palmar ligaments*, excepting in the *thumb*, which has a *capsular ligament* connecting it with the trapezium.

The *metacarpal bones* are united at their bases by *transverse dorsal*, and *palmar ligaments*, and by *interosseous ligaments*, which pass between their contiguous surfaces; these may be demonstrated by tearing the bones apart when the dissection is completed.

The METACARPO-PHALANGEAL ARTICULATIONS are maintained by *anterior* and *lateral ligaments*, and by *transverse ligaments*, which hold together the heads of the metacarpal bones of the four fingers. The extensor tendon of the finger takes the place of a posterior ligament.

The PHALANGEAL ARTICULATIONS have three ligaments; the *anterior*, firm and fibro-cartilaginous, and grooved for the flexor tendons; the *lateral ligaments*, one on each side, triangular in form, the apex being attached to the phalanx in front, and the base to the tubercle at the side of the phalanx behind. The extensor tendons supply the place of posterior ligaments.

14*

PART THIRD.

ANATOMY OF THE ABDOMEN AND LOWER EXTREMITY.

DISSECTION I.

PARIETES OF THE ABDOMEN.

A block is to be placed under the lumbar vertebræ, and the abdominal muscles made tense by inflating the peritoneal cavity by a blowpipe introduced through the umbilicus. A longitudinal incision is to be made from the ensiform cartilage to the pubes, penetrating to the tendons of the muscles; the amount of fat to be divided varies so much that great caution is necessary lest the tendons themselves be wounded; when reached they are known by their white and glistening aspect. From just below the umbilicus a second incision is to be carried upward and outward to the most dependent part of the margin of the thorax. The precaution above mentioned must also be observed in the inner half of this incision, which posteriorly should penetrate to the muscular fibres. The two angular flaps of integument thus formed are to be reflected upward and downward. The dissector must keep close to the tendon, cleaning it carefully and slowly from all the cellular tissue and superjacent parts, and, when dissecting the muscular portion, follow the direction of its fibres, freeing them patiently, one by one, from their sheath, being particularly careful at the point of their junction with the broad flat tendon into which they are inserted, not to divide and dissect up the tendon itself, but to keep it intact in its whole extent. Toward the groin more or less of the fascia should be left, in order to observe its relations to the external abdominal ring. The external oblique muscle cannot be exposed posteriorly without turning the subject over upon its face, or at least upon its side. The abdominal muscles are almost invariably discolored of a greenish hue; this is not usually owing to decomposition, but to the effect of sulphuretted hydrogen in the intestinal canal upon the coloring matter of the blood in the muscular tissue.

The PARIETES OF THE ABDOMEN extend from the median line to the spinal column on each side, and from the ribs above to the pelvis below; they include three pairs of flat muscles, disposed in layers, and the direction of whose fibres is different; in front they terminate in extensive aponeuroses, also disposed in layers, and between them on each side of the median line extends a single long muscle. The abdomen is covered by a fascia and a very variable amount of fat.

The *superficial fascia* is continuous with that of the thorax and lower extremity. No special dissection of it is necessary. It is only important in the inguinal region, where it divides into two layers, and, as will hereafter be seen, bears certain relations to the surgical affection called hernia; these two layers are separated by the *superficial epigastric artery* and *vein*, the former being a small branch from the femoral, arising below Poupart's ligament, and passing upward toward the umbilicus. It will be seen that certain fibres of the superficial fascia pass from the pubes to the penis, forming a rounded cord called the *ligamentum suspensorium penis*. In the groin the fascia attaches itself to Poupart's ligament, and through that becomes blended with the fascia of the thigh; it covers the spermatic cord, and, accompanying it to the scrotum, there unites with the fascia of the perineum. Between its two layers, and above Poupart's ligament, three or four lymphatic glands will be found; these receive the lymphatics from the abdomen and genital organs, and their efferent ducts pass in at the saphenous opening of the thigh.

The OBLIQUUS EXTERNUS MUSCLE is the most superficial of the abdominal muscles; it is aponeurotic in front and fleshy upon the side; it arises from the external surface of eight or nine lower ribs by processes called digitations, which are received between similar processes belonging to the serratus magnus and latissimus dorsi. It is inserted into the outer edge of the anterior half of the crest of the ilium and its anterior superior spinous process, into the spine, pectineal line and front of the os pubis, and into the whole length of the linea alba. The portion inserted into the crest of the ilium is fleshy; the remainder consists of a spreading aponeurosis, connected above with the pectoralis major muscle, and along the median line uniting with its fellow of the other side; the interlacement of their fibres forms what is called the *linea alba*, in the centre of which is the umbilicus, being the cicatrix of the occluded extremities of the umbilical artery and vein which composed the umbilical cord, divided at the time of birth, and the stump of which subsequently sloughs off. Between the anterior superior spinous process of the ilium and the spine of the pubes the fibres of the external oblique become rolled into a sort of cord, known as *Poupart's ligament*, and which is continuous by its lower border with the fascia lata of the thigh. When properly dissected, small foramina will be noticed in the aponeurosis, giving exit to the cutaneous nerves and vessels; the nerves are branches of the intercostals, and of the lumbar plexus, and the arteries are chiefly from the internal mammary, lumbar, and circumflexa ilii. The outline of the rectus muscle can be plainly seen through the tendon of the external oblique; the curved line which indicates its external border and extends from the os pubes to the chest is called the *linea arcuata*. At the pubes the fibres of the aponeurosis split, and leave a space which affords passage to the spermatic cord in the male and the round ligament in the female: this space is very variable in size, and is usually larger in the male than the female subject; although there is nothing annular in its conformation to give it such a name, it is called the *external abdominal ring*. The fibres forming its superior border or *pillar* interlace with those of the opposite side in front of the symphysis pubes; its inferior border or pillar is formed from the internal portion of Poupart's ligament. Just above the spermatic cord this separation in the aponeurosis is traversed by a series of transverse fibres, variable both in size and number; they extend for a considerable distance on either side of the pillars, and constitute what are called the *inter-columnar fibres*. Between the pillars, covering in the cord and prolonged upon it, is a thin and delicate expansion called the *spermatic fascia;* if this is divided transversely upon the cord, the handle of the scalpel may be passed under it and pushed upward beneath the tendon; this will demonstrate its existence as a layer distinct from the elements of the cord.

The external oblique is to be removed by dividing it transversely across its fleshy part; the change in the direction of the fibres will show when the internal oblique is reached. In accomplishing the separation of these two muscles, the fascia of the internal muscle should be removed with the external. In order to reserve the inguinal region, the tendon should be divided transversely from the anterior superior spinous process to the linea alba, and down the linea alba to near the pubes, so far as the close union with the muscle beneath will permit, the portion below being left for further examination. The obliquus internus is thus exposed, except at its lower part, which, by a little manipulation, may also be seen, on turning downward that portion of the external oblique which has been left attached.

The OBLIQUUS INTERNUS MUSCLE arises from the outer half of Poupart's ligament, from the anterior two-thirds of the crest of the ilium, and from that portion of the lumbar fascia which is attached to the spinous processes of the lumbar vertebræ. The fibres of the lower part of this muscle are thin, and separated from each other; they curve over the spermatic cord, or round ligament, and uniting with the tendon of the muscle beneath, under the name of the *conjoined tendon of the internal oblique and transversalis muscles*, are inserted into the crest of the pubes and the pectineal line behind the tendon of the external oblique. The remainder of the fibres pass upward and inward, and terminate in an aponeurosis at the outer border of the rectus muscle. The upper half of this aponeurosis splits into two laminæ, which encase the upper half of that muscle, and meet on the median line; the whole of the aponeurosis passes in front of the lower half of the muscle, blending with the tendon of the external oblique. Along the linea alba, the tendons of the two sides unite inseparably. Superiorly the muscle is inserted into the lower border of the cartilages of the last four ribs.

In arching over the spermatic cord, certain of the muscular fibres are prolonged, and carried downward, in long loops, by the testicles at the period of their descent; these fibres may be seen as a muscular layer upon the cord, and are called the *cremaster muscle;* they vary very much in distinctness, and do not exist in the female subject. They are sometimes described as coming from the transversalis muscle, as well as the internal oblique, the two muscles being intimately blended at this point.

The separation of this muscle from the next is with difficulty accomplished in a manner which leaves the transversalis neatly and fairly exposed. The fibres of the internal oblique should be divided transversely to their direction; the different direction of its fibres and the ramifications of vessels in the space between the muscles, will tell when the transversalis is reached; the internal oblique is to be dissected away from it, and divided from its own tendon along the outer edge of the rectus muscle. The lower fibres, which assume a similar direction to those of the transversalis, and to a certain extent become confounded with them, may be left behind, with reference to a more special examination of the inguinal region in relation to hernia.

The TRANSVERSALIS MUSCLE arises from the outer half of Poupart's ligament, from the anterior three-fourths of the inner lip of the crest of the ilium, from the lumbar fascia, *i. e.*, from the spinous processes and the tips and bases of the transverse processes of the lumbar vertebræ, and from the under surface of the last six or seven ribs, where it indigitates with the diaphragm. Its fibres pass transversely forward, and terminate in an aponeurosis, the upper two-thirds of which passes behind, and the lower third in front of the rectus muscle, to be inserted into the linea alba. The lower fibres of the muscle cover over the spermatic cord, or round ligament, and are inserted into the pectineal line, in connection with the lower fibres of the internal oblique, under the name of the *conjoined tendon :* these lower fibres are few in number, and separated from each other; indeed, the portion arising from Poupart's ligament is sometimes deficient, and at others confounded with, and inseparable from, the internal oblique; in either case the transversalis fascia and peritoneum may be seen through the fibres: the internal oblique usually descends nearest to Poupart's ligament, while the transversalis makes up the greater part of the conjoined tendon. The circumflexa ilii artery, a branch of the external iliac, and the musculo-cutaneous nerve from the lumbar plexus, ramify upon and in this muscle above the crest of the ilium.

The rectus muscle is exposed by dividing the aponeurosis which covers it in front, from the sternum to the pubes, at about an inch from the median line; at two or three points this sheath will be found adherent to certain tendinous intersections which traverse the muscle; the aponeurosis should be carefully dissected from them.

The RECTUS MUSCLE, broad above and narrow below, arises by a thick tendon from the crest of the os pubis; becoming thinner as it grows broader, it is inserted into the cartilages of the fifth, sixth, and seventh ribs; it is separated from its fellow by an interval, formed by the aggregation of the fibres of the tendons of the abdominal muscles of the two sides, which constitutes the linea alba. Two or three tendinous intersections, sometimes complete, at others only partial, cross the rectus in irregular directions; they are called *lineæ transversæ*, and serve to keep the muscle flat during its contractions, since without them, its broad insertion and narrow origin would give it a tendency to roll into a cone longitudinally. The rectus muscle is occasionally inserted as high as the fourth, or even third rib. Sometimes there may be found an abnormal muscle upon the sternum, which appears, physiologically at least, to be a continuation of this muscle; it is called the *rectus sternalis*, and is described as being occasionally connected with the sterno-mastoid, instances being recorded where the sterno-mastoid and rectus were one continuous muscle. This occasional slip, and the high insertion of the rectus, is looked upon as corresponding with the long rectus of the lower orders of animals.

The PYRAMIDALIS MUSCLE, small and triangular in shape, lies upon and inclosed within the sheath of the rectus, and is, as it were, accessory to that muscle, which becomes enlarged at its lower part when it is absent; it arises from the crest of the os pubis, and, tapering as it goes upward, is inserted into the linea alba from two to four inches above the symphysis pubes. This muscle is very commonly wanting on one or both sides.

The rectus muscle should be divided at the umbilicus; the two ends will be easily dissected up from their posterior cellular attachments, and it will then be seen how it is encased by the aponeurosis of the abdominal muscles.

In front of the upper half of the rectus muscle expands the aponeurotic tendon of the external oblique, and the anterior layer of that of the internal oblique; behind the upper half are the posterior layer of the tendon of the internal oblique, and the whole of that of the transversalis. Below the umbilicus all the tendons pass in front, while behind, the rectus is separated from the viscera only by the transversalis fascia and the peritoneum. The lower edge of the tendons forming the posterior sheath of the upper half of the rectus muscle assumes a crescentic shape, and is called the *semilunar fold of Douglass;* the space below this fold, between the peritoneum and the muscle, Retzius has described as a "*pre-peritoneal cavity*," having proper walls, formed by the transversalis fascia, and destined to accommodate the bladder in its changes of volume, and enabling the recti muscles to act more directly upon that viscus in emptying it of its contents. Below the umbilicus the linea alba becomes more tendinous than above that point, and the peritoneum is manifestly less closely connected with the abdominal walls. The *epigastric artery*, a branch of the external iliac, will be seen penetrating the lower half of the rectus, and lying loose between it and its sheath for a portion of its course; if well injected, it may also be traced in the fibres of the muscle to an anastomosis with the terminal branches of the internal mammary artery. It was by the aid of these anastomoses that some of the old anatomists endeavored to explain the intimate relationship between the genital organs and the mammary gland. At the lower part of the inner surface of the rectus muscle will be seen a thin layer of diverging fibres, springing from the insertion of Poupart's ligament into the pectineal line, and ascending upward and inward to the linea alba in which they are lost; it is called the *triangular ligament*.

ANATOMY OF INGUINAL HERNIA.

It is always desirable, if possible, to make a special dissection of the inguinal region; but if the student has left the lower part of the abdominal muscles as directed, it may be very satisfactorily examined after the general dissection is accomplished. The dissection can be advantageously performed only upon the male subject.

Under the influence of violent exertions, and sometimes passively, a portion of the contents of the abdomen may be protruded at such parts of its walls, as, from their conformation, are weaker or less protected than others. This protrusion is called hernia, or rupture. The inguinal region is a point at which it frequently occurs. To understand why this region is liable to this accident, it is necessary to examine it specially in this relation.

The INGUINAL REGION is included between Poupart's ligament, the linea alba, and an imaginary transverse line from the latter to the anterior superior spinous process of the ilium; most of its parts have already been separately described. The perforation of the external oblique by the spermatic cord, the thinness, or partial deficiency of the fibres of the internal oblique and transversalis muscles, make it apparent that less resistance would be offered here than elsewhere to the superincumbent weight of the abdominal viscera, and that their impulsion against it, during efforts which contracted the diaphragm and abdominal muscles, would render protrusion of the intestines or omentum at this point a very conceivable occurrence.

The *superficial fascia*, but for its relation to the anatomy of hernia, would be considered merely as the sheath of the external oblique, in which a certain amount of fat was deposited. It is usually spoken of as consisting of two layers, separated by the superficial epigastric artery and vein. Unless a special dissection of it has been made, it will probably be found that one of these layers has been removed with the skin. Directly beneath this is the tendon of the external oblique, between the fibres of which emerges the spermatic cord. This point of emergence is called the external abdominal ring, and consists merely of a separation of the tendon, the upper border of which is called the *superior pillar*, and the lower, the *inferior pillar* of the ring. The further separation of these pillars is prevented by transverse fibres, called *inter-columnar*. The *spermatic cord* consists of the excretory duct of the testicle, called the vas deferens, the spermatic artery and vein, a nerve, some lymphatics, and cellular tissue. These constitute a bundle of considerable size, reinforced by the cremaster muscle, which is not properly a part of the cord, the whole of which lies behind a thin fascia, extending across the pillars of the external ring, and continued down upon the cord, called the *spermatic fascia*.

Upon reflecting the lower part of the external oblique, toward the groin, the spermatic cord will be observed lying for a short distance behind it; the internal oblique and transversalis curve over it, and Poupart's ligament is directly beneath it. The cremaster muscle, being fibres of the internal oblique and transversalis muscles, lies upon the cord itself, and its connection with those muscles may possibly be traced. The space traversed by the cord, behind the external oblique, is called the *inguinal canal*.

If the internal oblique and transversalis are separated, upon turning down the lower part of the former muscle, it will be seen that the transversalis leaves a space between itself and Poupart's ligament; in this space will be seen the *transversalis fascia*, and upon reflecting the transversalis muscle, it will be seen lying behind it, upon the peritoneum. This, like the superficial fascia, were it not for its relations to hernia, would be considered merely as the sheath of the transversalis muscle; it is of very variable thickness, sometimes amounting to nothing but a little cellular tissue; it is attached to Poupart's ligament below, and internally to the conjoined tendon and sheath of the rectus muscle. By pulling the spermatic cord, it will be seen that this fascia is reflected from the peritoneum on to the surface of the cord, and the cone which is thus formed and made apparent, by this traction, is called the *infundibuliform fascia*. The orifice forming the base of this cone, and across which the peritoneum is stretched, is the *inter-*

nal abdominal ring. By incising the fascia on the peritoneum, the handle of a scalpel may be inserted, and pushed down the cord between it and that portion of the transversalis fascia just described as the infundibuliform, showing that at this point the peritoneum might be protruded before a knuckle of intestine, and, if forced onward, must pass down in the direction taken by the knife-handle, between the fascia and the cord. The epigastric artery lies between the transversalis fascia and the peritoneum; it passes under the cord close to the inguinal ring; the vas deferens hooks over it as it turns downward into the pelvis.

A large triangular flap of the peritoneum being incised and turned toward the groin, it will be noticed that, upon the median line, there is a cord passing from the bladder to the linea alba; this is the remains of the allantois of fœtal life; another cord, being the remains of the obliterated hypogastric artery of fœtal life, passes obliquely from the umbilicus downward toward the pelvis, in a line nearly corresponding to that of the epigastric artery: this cord, by its shortness, causes the peritoneum to make a pouch on each side—these are called the *inguinal fossæ;* one of them is behind the external ring, and the other behind the internal ring, and it will be seen that each must direct that portion of intestine which accidentally lies within it toward the ring that lies in front of it, and as the abdominal wall corresponding to the external ring is weaker than elsewhere, owing to the loss, at that point, of one of the layers constituting its thickness, and at the internal ring, owing to the conformation already described, it follows that a hernia is liable to occur at either situation; it is called a *direct hernia,* if it protrudes from directly opposite the external ring, and an *oblique hernia,* if it enters the internal ring, and follows the direction of the spermatic cord.

By pulling upon the spermatic cord, the peritoneum will exhibit, at the part affected by the traction, a puckered appearance; this is the point at which that point of the peritoneum, carried before the testicle in its descent to the scrotum, is obliterated from its connection with the general cavity. The portion of peritoneum intervening between this point and the scrotum, usually degenerates into cellular tissue; it may, however, remain as a distinct cord, or even as a pervious tube. In the female subject, the round ligament enters the internal ring in place of the spermatic cord, and also carries before it a pouch of peritoneum, but only for a short distance; the diverticulum, which the peritoneum forms, is called the *canal of Nuck*.

DIRECT INGUINAL HERNIA commences in the internal inguinal fossa, in a triangle, called the *triangle of Hesselbach*, formed by the external edge of the rectus muscle, Poupart's ligament, and the obliterated hypogastric artery; it carries before it the peritoneum which forms the *sac* of the hernia, the transversalis fascia, the conjoined tendon (this, however, it sometimes splits, and passes through, instead of pushing before it), the spermatic fascia, the superficial fascia, and the skin; it then descends toward the scrotum. The epigastric artery and the spermatic cord are both left upon the outer side of this form of hernia.

OBLIQUE INGUINAL HERNIA commences in the external inguinal fossa, and from that derives its peritoneal sac; the internal inguinal ring entered, it pushes before it the infundibuliform fascia, and when in the canal, the cremaster muscle; at the external ring, the spermatic fascia, the superficial fascia, and the skin; it then descends toward the scrotum. In this form of hernia, the spermatic cord is beneath, and behind the tumor formed by it; the epigastric artery is upon the inside, the reverse of its position in direct hernia. Oblique hernia, from its size and long existence, may so enlarge and drag upon the internal ring, as to bring it behind the external ring, and thus assume the appearance of a hernia primarily direct. The artery, of course, retains its relative position.

When the portion of peritoneum, carried before the testicle in its descent, has not been obliterated, and the intestine passes through the tube thus left, into the tunica vaginalis testis, it is called a *congenital hernia*. When this tube is partially obliterated, and admits the intestine into its upper part only, it pushes down behind that portion which has formed the tunica vaginalis testis, and is called an *encysted*, or *infantile hernia*. Inguinal hernia rarely occurs in the female, but all the various forms have been noticed; in females the hernia descends into the labia pudendi.

DISSECTION II.

VISCERAL CAVITY.

In opening the abdomen, a thin peritoneal lamina, extending from the umbilicus to the liver, will be seen; this is the *broad ligament* of the liver; in the free border of this, is a round cord of considerable size, called the *round ligament*, being the obliterated umbilical vein of fœtal life. These should be examined, at this time, as so good an idea of them cannot be obtained after their division. Hernia may occur at the umbilicus, especially during infancy, before the parts have consolidated. Its course is direct, and it has for its coverings the integument, superficial fascia, a prolongation from the tendinous margin of the abdominal opening, the fascia transversalis, and peritoneum.

The abdomen is now to be opened, by incising longitudinally and transversely, whatever of the anterior parietes remains, and reflecting the flaps thus made.

The VISCERAL CAVITY being exposed, it will be remarked that the intestinal tube, with its various divisions and convolutions, occupies, apparently, the whole of the abdomen, the only viscus attracting attention at the first glance being the liver. The intestines will be found partially covered, superficially, with a membranous flap, or apron; at times this is crumpled up (so to speak), and lies a confused mass, occupying but a small space; it may, however, be spread out; it is called the *great omentum*. Occupying the upper part of the cavity, and in close apposition to the diaphragm, will be seen the liver; partly covered by the liver, and filling the left side, is the stomach, behind which may be found the spleen. Below the stomach, and stretching across the spine, is the pancreas. From the stomach may be traced the small intestine; near the union of this with the stomach the surrounding parts will often be found discolored by bile which, after death, transudes through the walls of the gall-bladder. The small intestine occupies the central part of the visceral cavity and the pelvis; by running it through, from its commencement, it will be found that it joins the large intestine, or colon, in the right inguinal region. The large intestine ascends on the right side, covering in the kidney, and is called the *ascending colon;* it crosses the abdomen at its upper part, under the name of the *transverse colon*, descends upon the left side, as the *descending colon*, covering in the left kidney, to the left inguinal region, where it forms several folds, called the *sigmoid flexure*, and then dipping into the pelvis behind the bladder becomes the *rectum*.

All the viscera, so far as they can be seen, as well as the walls of the abdomen, will be found covered with a thin, shining membrane, called the peritoneum; this facilitates the movements of the various organs, which are by no means inconsiderable, the position of the body influencing very greatly the position of the viscera. They have, notwithstanding, been located in regions, and though the boundaries are merely arbitrary, and are indicated only in the most general way as including the organs within their limits, allusion is constantly made to them.

It will be found convenient to indicate these regions by strings, which should be stretched in the directions about to be enumerated, viz.: two vertical lines, each from the most dependent portions of the cartilages of the eighth ribs to the centre of Poupart's ligament; a transverse line corresponding to the summits of the ilia, and another to the most dependent portion of the ribs.

We thus have three zones, each subdivided into three regions. The three in the upper zone are, laterally, the right and left *hypochondriac*, in the centre, the *epigastric*. In the middle zone, laterally, the right and left *lumbar*, and in the centre the *umbilical*. In the lower zone, laterally, the right and left *inguinal*, and in the centre the *hypogastric*. By handling the abdominal contents, the student can satisfy himself that the right hypochondriac region contains the liver, the left the spleen, and part of the stomach, while the epigastric contains a part of both the stomach and the liver; the lumbar regions contain the kidneys on either side; the umbilical the small intestines; the right inguinal region contains the cæcum, the left the sigmoid flexure; the hypogastric region contains the bladder and rectum, and in the female, the uterus.

PERITONEUM.

The PERITONEUM is a serous membrane. Considered in its simplest form, a serous membrane is a hollow sac, into which a viscus protrudes itself, and thus, while getting a covering itself, lies within another covering of the same membrane, reflected from its own walls. We have only to conceive this sac as large enough to admit of many viscera protruding into it, and we have an idea of the peritoneum; some of the viscera it merely passes over, without investing them on all their sides, and all of them are, in reality, outside the membrane. The folds made by the peritoneum, either in its reflections from the viscera which it has invested, or in passing from one organ to another, constitute means of support, and hold them in their proper places; these in certain instances are improperly, though very conveniently, called *ligaments*, in others *omenta*, and in still other instances *mesenteries;* besides supporting the organs, they contain the various vessels and nerves destined to the different parts.

The *great omentum* is formed by two folds of the peritoneum, which descend from the larger curvature of the stomach, covering in the small intestines, then, doubling upon themselves, they return to be attached to the transverse colon ; it therefore consists of four thicknesses ; they are not, however, easily separable. The great omentum contains a variable amount of fat, is sometimes perforated with holes, and is not unfrequently gathered up in a mass near the stomach.

The *lesser omentum* is the fold extending between the smaller curvature of the stomach and the liver; it contains the portal vein, hepatic artery and ducts; its left side is continuous with the œsophagus, but its right forms a free margin, beneath which the finger may be passed into the cavity of the peritoneum behind the stomach; this passage is called the *foramen of Winslow*.

The *gastro-splenic omentum* passes from the outer border of the stomach to the spleen, and contains the splenic vessels.

The *mesentery* proper is that fold of the peritoneum which holds down the small intestine, and which is attached posteriorly to the front of the spine ; it is about four inches wide, and between its two layers, besides a considerable amount of fat, are the arteries, veins, and nerves of the small intestine, also the lymphatic vessels and glands, called the mesenteric glands.

The *transverse meso-colon* is the mesentery of the transverse colon, and its medium of connection with the posterior wall of the abdomen.

The *meso-rectum* and *meso-cæcum* are similar folds connected with the rectum and cæcum. The meso-cæcum is of such length as to permit great mobility to the cæcum, though it is usually described as closely bound to the iliac fossa. A left inguinal hernia of the cæcum may occur.

The complications of the peritoneum are not easily comprehended, and it is only after a good deal of thought and repeated examination that they can be fully understood. In tracing the continuity of this membrane from above downward, the student begins at the liver, where he will perceive that it is prolonged from the under surface of that viscus on to the vessels. "From the liver it may be followed along the vessels, one layer before and the other behind them, forming the lesser omentum, to the upper border of the stomach. At the stomach the two layers inclosing the vessels separate, one going before and the other behind it; but beyond that viscus they are applied to one another to form the great omentum. After descending in contact in that fold to the lower part of the abdomen, they may be traced in it backward and upward, and may be seen to separate to inclose the transverse colon, like the stomach, and then to continue to the spine, giving rise to the transverse meso-colon. At the attachment to the spine the two companion layers will be found to separate, one passing upward and the other downward. The ascending layer is continued in front of the pancreas and the pillars of the diaphragm, and blends with the peritoneum on the posterior aspect of the liver. The descending layer may be followed from the transverse meso-colon along the middle line of the spine, over the duodenum, the aorta, and vena cava, till it meets with the artery to the small intestine, along which it is continued to form the mesentery, turning over the intestine and back to the spine along the other aspect of the vessels. From the root of the mesenteric artery the peritoneum descends to the pelvis, and partly covers the viscera in that cavity; thus it surrounds the upper part of the rectum, and attaches this viscus to the abdominal wall by the meso-rectum; next, it is continued forward between the rectum and the bladder, or between the rectum and the uterus, where it forms a pouch; thence it passes over the back and sides of the bladder. Lastly, the serous membrane is continued to the inguinal region, where it forms the fossæ, before alluded to (p. 170), and it can be traced upward on the anterior wall of the abdomen and the diaphragm, to the rest of the membrane on the upper surface of the liver."[1]

The peritoneum may also be traced in a transverse direction. Beginning at the umbilicus, it may be followed outward to the large intestine, which it fixes to the abdominal wall by the meso-colon, over the kidney to the middle line, along the vessels to the small intestine, round the intestine and back to the spine along the other aspect of the

---

[1] Ellis's Demonstrations of Anatomy.

vessels, and so outward over the large intestine of the opposite side, to the parietes of the abdomen and the umbilicus.

DUCTS, VESSELS, AND NERVES OF THE ABDOMINAL CAVITY.

By raising the edge of the liver and drawing the intestines downward, the *biliary ducts*, situated in the lesser omentum and lying between the two peritoneal layers forming the pillar of the foramen of Winslow, may be examined.

From the duodenum a duct of variable size can be traced backward toward the liver; this is the *ductus choledochus communis;* it divides as it approaches the liver into two branches, one of which comes from the neck of the gallbladder, and is called the *cystic duct*, the other, coming from the transverse fissure of the liver, the *hepatic duct;* they are accompanied by the hepatic artery and portal vein.

The veins of the portal system may be found by pushing aside the viscera and searching for them in the regions to which they belong; it is impossible to dissect them without making special preparation for so doing.

The PORTAL VEIN or SYSTEM is composed of those vessels which return the blood from the chylopoietic viscera; they are the

Inferior Mesenteric,	Splenic,
Superior Mesenteric,	Gastric.

The *inferior mesenteric vein* returns the blood from the rectum, sigmoid flexure and ascending colon, terminating in the splenic vein.

The *superior mesenteric vein* returns the blood distributed by the superior mesenteric artery; it ascends in company with that vessel, and behind the pancreas unites with the splenic vein.

The *splenic vein* commences in the spleen by several branches, which, uniting, form a large trunk which passes behind the pancreas, receiving the *gastric veins* from the stomach and duodenum. The union of this with the superior and inferior mesenteric veins forms the portal vein.

The *portal vein* lies in the lesser omentum, between the biliary ducts and hepatic artery; it ascends to the transverse fissure of the liver, where it divides into two branches, one for the right and the other for the left lobe.

The arteries of the various abdominal viscera all come from the abdominal aorta; the nerves from the pneumogastric and from the sympathetic ganglia. Their dissection is one of considerable difficulty, owing to the mobility of the viscera and the constant necessity of changing their position, so as successfully to expose the different trunks. The aorta should first be found at the point where it perforates the diaphragm, then, by removing the peritoneum, each artery, as it presents itself, is to be followed out by such means as the dissector's ingenuity will suggest. The nerves will some of them be exposed with the arterial trunks which they accompany; a detailed dissection of them is rarely accomplished: being composed of plexuses proper to each organ, they are usually made the subject of special dissections. More than to trace, to some slight extent, the connection of these plexuses with the principal ganglia of the sympathetic, can hardly be expected. The student, therefore, need not be disappointed if he is unable to verify those paragraphs marked with an asterisk.

The nervous plexuses of the abdomen may be enumerated as follows:—

Solar,
Phrenic,
Supra-renal,
Gastric,
Hepatic,
Splenic,
Superior Mesenteric,
Aortic,
Inferior Mesenteric,
Renal,
Spermatic,
Hypogastric.

In searching for the cœliac axis, which lies behind the stomach, the SOLAR PLEXUS will be seen encircling the cœliac axis, covering the aorta, and spreading out in all directions; it is joined by the greater splanchnic nerve of both sides (p. 122), and receives branches from the pneumogastric and phrenic nerves. The solar plexus contains a number of ganglia, the principal one of which is called the SEMILUNAR GANGLION, and the radiation of nerves from this gives the plexus its name of solar. From this plexus arise those branches which form the *phrenic* and *supra-renal plexuses;* these may also be seen at this stage of the dissection.

The ABDOMINAL AORTA commences at the aortic opening of the diaphragm, and, descending on the left side of the vertebral column, terminates by dividing into the two common iliac arteries.

The CŒLIAC AXIS is the first large trunk given off by the abdominal aorta. The *phrenic arteries* are normally the first, but they are extremely irregular in their origin, and as often arise from the cœliac axis as from the aorta; they are two in number, and ascend obliquely outward, ramifying on the under surface of the diaphragm; their inosculations with other arteries are very numerous, and they send a branch to the supra-renal capsule on both sides. The cœliac axis is a short but large trunk, which arises close to the diaphragm and quickly divides into three branches, the gastric, hepatic, and splenic. Each of these branches conveys the plexus of nerves destined for the organ to which it is distributed, and it may be seen surrounding the artery as a sort of sheath. The cœliac axis may be wanting, its branches originating directly from the aorta; or it may give off but two branches, the gastric and the splenic, the hepatic coming from some other source, as, for instance, the superior mesenteric or the aorta.

The *gastric artery* is distributed to the smaller curvature of the stomach; it joins it near the œsophagus, to which it gives some ascending branches, while others pass round the cardiac extremity to join the vasa brevia of the splenic artery; its terminal branch unites with the pyloric branch of the hepatic.

The *hepatic artery* curves up toward the liver, and is in close relation with the hepatic duct and portal vein; near the pylorus it gives off a *pyloric* branch to the lesser curvature of the stomach, where it inosculates with the gastric, and also the *gastro-duodenalis*, which supplies the stomach and duodenum; the gastro-duodenalis divides into the *gastro-epiploica dextra*, which is distributed to the larger curvature of the stomach and anastomoses with the splenic, and into the *pancreatico-duodenalis*, supplying the duodenum and pancreas, and anastomosing with the superior mesenteric. The hepatic at its termination divides into two branches which enter the transverse fissure of the liver and separate to its right and left lobes. A small branch, the *cystic*, is given off by one of these to the gall-bladder. *Accessory hepatic arteries* are frequently found, usually coming from the gastric artery.

The *splenic artery* is the largest branch of the cœliac axis; it supplies the spleen, pancreas, and stomach; it passes outward in a tortuous manner, behind the pancreas, and divides into numerous terminal branches which penetrate the hilus of the spleen; the splenic vein is in close relationship with it. It reflects several small branches to the great end of the stomach which are called *vasa brevia*, and a single large branch passes along the greater curvature of the stomach, under the name of *gastro-epiploica sinistra*, anastomosing with a branch from the hepatic. Numerous small twigs are given off to the pancreas.

The SUPERIOR MESENTERIC ARTERY arises from the front of the aorta just below the cœliac axis, sometimes in connection with that trunk. A portion of this artery at its commencement is covered in by the pancreas, to which it sends a small branch; it divides into numerous branches, which, placed parallel to each other, descend between the two layers of the mesentery, and then by a series of vascular arches, rarely if ever exceeding three in number, supplies the small intestine from the duodenum to its termination in the colon; these arches anastomose so freely with each other, that the circulation at any one point can never be interrupted by the compression of the intestinal folds. Besides supplying the small intestine, the mesenteric artery sends a separate division, called the *ileo-colic*, to the cæcum and termination of the ileum; another single division supplies the ascending colon, and is called the *colica dextra;* a third branch supplies the transverse colon, being called the *colica media;* these three trunks anastomose together by their terminal branches, and the last named inosculates with the colica sinistra, a branch of the inferior mesenteric, which supplies the descending colon and sigmoid flexure, "and so completes," says John Bell, "the great mesenteric arch, one of the most celebrated inosculations in the whole body, that of the circle of Willis hardly excepted."

* These several branches of the mesenteric artery are accompanied by the nerves which constitute the *superior mesenteric plexus;* they can be found, together with several ganglionic masses, at the commencement of the artery, which they surround, and may be traced back to the solar plexus; they supply the pancreas and intestinal canal.

In the intervals of the branches of the mesenteric vessels numerous lymphatic glands are lodged; these are called the *mesenteric glands;* the chyliferous vessels of the small intestine pass through these to reach the thoracic duct; these glands are often enlarged by disease.

The INFERIOR MESENTERIC ARTERY arises about two inches below the superior, also from the front of the aorta; it inclines to the left, and by a series of arches supplies the descending colon under the name of the *colica sinistra*, forming large and free anastomoses with the colica media of the superior mesenteric; a second branch passes to the sigmoid flexure, and is called the *sigmoid artery*, and its terminal branch goes to the mesentery of the rectum, under the name of the *superior hemorrhoidal artery;* these three branches anastomose freely with each other.

* The front of the aorta, between the mesenteric arteries, is covered by the *aortic plexus* of nerves; this connects with the solar plexus, and also unites with the lumbar ganglia, situated upon the sides of the vertebræ; from it passes off the *inferior mesenteric plexus*, which, accompanying the artery of that name, supplies the parts to which that vessel is distributed.

The intestines should now be removed. To do this the rectum is to be tied and divided; the large intestine should then be dissected from the meso-colon close to the intestinal wall, and the small intestine separated from the mesentery in a similar manner, tying and dividing it just below the duodenum; they are then to be set aside for further examination, and kept immersed in water.

The RENAL ARTERIES are given off at right angles from the sides of the aorta, just below the origin of the superior mesenteric artery; the right is given off lower down, and is shorter than the left; each divides into several terminal branches to enter the kidney at its hilus; occasionally they penetrate through the sides of the organ. The aorta sometimes gives off two, three, and even four, renal arteries to one kidney, and they often vary in arrangement on the two sides of the body.

* The *renal plexus* lies upon the renal artery, and is composed of branches from the solar and aortic plexuses; the inferior and renal splanchnic nerves (p. 122) also terminate in this plexus. Ganglia of various size occur in it, and its branches ramify upon and in the substance of the kidney.

A small branch from the side of the aorta just above the renal artery goes to the supra-renal capsule, and is called the SUPRA-RENAL ARTERY. The capsule usually receives a branch from the renal and also from the phrenic arteries.

The SPERMATIC ARTERIES are two long and slender branches given off from the front of the aorta, just below the origin of the renal arteries; occasionally they arise by a common trunk; sometimes there are two upon one side, and their origin may be from the renal artery. They pass downward beneath the peritoneum, crossing the ureter and external iliac artery, to the internal inguinal ring, and then accompany the spermatic cord to the testis. In the female these arteries are called the ovarian, and instead of passing out of the abdominal cavity, they dip into the pelvis, and pass up between the layers of the broad ligament of the uterus to the ovaries. The length of these vessels will be accounted for, when it is remembered that the testicle descends at the close of intra-uterine life to the scrotum from the lumbar region, and that the ovary, during pregnancy, is lifted with the uterus above the umbilicus; it is therefore for the purpose of accommodating these displacements of the organs to which they are distributed.

* The *spermatic plexus* accompanies the spermatic arteries; it is formed by branches from the renal and aortic plexuses.

The *spermatic veins* accompany the spermatic arteries in a part of their course; they are formed by the union of several venous branches which surround the spermatic cord, and ascend as single trunks, the right to enter the vena cava and the left to join the left renal vein. It has been attempted to explain the more frequent occurrence of varicocele upon the left side, by the less direct entrance of the blood from the vein of that side into the general circulation.

The abdominal aorta will be found to terminate in two branches at the level of the third lumbar vertebra, called the COMMON ILIAC ARTERIES; these diverge, and divide, opposite the union of the sacrum with the os innominatum, into *external* and *internal iliac* arteries; the former courses along the brim of the pelvis to pass out beneath Poupart's ligament, and become the artery of the lower extremity; the latter dips into the pelvis to supply the viscera it contains. In negroes and in old subjects, the common and external iliac arteries are often curved and tortuous. Lymphatic glands will be found along their course.

At the point of the aorta's bifurcation a small artery arises called the SACRA MEDIA; it descends on the middle line of the sacrum, giving off short branches upon either side in its course toward the coccyx, where it terminates. It sometimes arises from the left common iliac.

The *lumbar ganglia*, forming part of the continuous chain of the ganglia of the sympathetic, which extends from the head to the coccyx, should now be dissected; their connection with the anterior branches of the spinal nerves can also be traced. The ganglia will be found much larger in size than those in the thoracic region; they also lie closer together; upon the right side they are covered in by the vena cava, on the left they rest upon the vertebræ along the edge of the psoas muscle. The branches which communicate with the anterior spinal nerves are of considerable length and accompany the lumbar arteries, passing beneath the tendinous fibres by which the psoas muscle arises.

* These ganglia give origin to the hypogastric plexus, which lies upon the anterior surface of the sacrum and last lumbar vertebræ, and distributes its branches to the viscera of the pelvis.

The LUMBAR ARTERIES are four in number, and correspond to the intercostal branches of the thoracic aorta.

They arise from the posterior aspect of the aorta, those of the right side being the longest and covered in by the vena cava; passing over the bodies of the vertebræ they dip beneath the psoas muscle and divide into two branches, one of which goes to the spinal cord and spinal muscles, while the other continues its course forward to the abdominal muscles. The first lumbar artery passes under the pillar of the diaphragm, and along the edge of the last rib; the last lumbar passes along the crest of the ilium. The lumbar arteries of the opposite sides sometimes arise by a common trunk, which sends out its branches laterally; two arteries of the same side are sometimes conjoined at their origin.

The EXTERNAL ILIAC VEIN is the continuation of the femoral vein, and will be found accompanying the external iliac artery; at Poupart's ligament it lies on the inside of the artery, but gradually gets beneath it; it is joined by the *internal iliac vein*, which returns the blood from the penis and pelvic viscera. These two veins form the *common iliac veins*, which, uniting upon the right side of the aorta, form the VENA CAVA INFERIOR; this ascends along the right side of the abdominal aorta, receiving in its course the *lumbar veins*, which accompany the lumbar arteries, the left being the longest; the *renal*, which lie in front of the renal arteries, the left renal crossing the aorta and being the longest; the *supra-renal*, which, however, sometimes terminate in the renal; the *phrenic*, returning the blood from the diaphragm, and the *hepatic*, which empties the blood from the portal system and the hepatic artery. The phrenic and hepatic veins cannot be seen until the liver is removed. The *spermatic veins* are long and slender veins accompanying the spermatic arteries; the right spermatic enters the front of the vena cava, the left spermatic enters the left renal vein; these return the blood from the scrotum and testicles. The vena cava passes out of sight beneath the liver, where it traverses a groove in that organ destined for it, and passing through a special opening in the diaphragm, penetrates the pericardium and terminates in the right auricle of the heart. The *vena azygos major* and *minor* (p. 123), originate from the lumbar veins, and pass up beneath the diaphragm to unite with the vena cava superior.

DISSECTION III.

INTESTINAL TUBE.

The intestines must be washed before they can be examined, but it should never be done until they are opened; this may be accomplished by the scissors, or by a special instrument called an enterotome; the section should be made along the line of attachment of the mesentery.

The portion of the alimentary canal known as the INTESTINES, extends from the stomach to the anus, and is divided into the small and large intestine. The *small intestine* extends from the pylorus to the cæcum, the *large* from the cæcum to the anus; they differ from each other in structure as well as size, the first being of uniform calibre, and the second much larger, and of a sacculated and irregular outline.

The small intestine, about twenty feet in length, is arbitrarily divided into three portions, duodenum, jejunum, and ileum. The *duodenum* is the first twelve inches beyond the stomach; the *jejunum* is the first two fifths of the remaining portion; the lower three fifths constitute the *ileum*, which terminates in the cæcum.

The large intestine, or *colon*, about five feet in length, is divided into the *cæcum*, or *caput coli*, the *ascending*, *transverse*, and *descending colon*, the *sigmoid flexure*, and the *rectum*. Attached to the cæcum is a small, tapering, tubular appendage, a little bulbous at its occluded extremity, from three to six inches in length, and held in its place by a mesentery proper to itself; it is usually tortuous, and about the size of a pipe-stem; it communicates with the interior of the cæcum through a small orifice, and is called the *appendix vermiformis cæci*. Two or three longitudinal bands run the whole length of the large intestine; they are about half an inch wide, and, being shorter than the canal itself, give to its walls a sacculated character.

To form a proper idea of the ileo-cæcal valve, the cæcum and a portion of the ileum should be inflated and dried; then cutting a window in the side, its arrangement may be seen.

The entrance of the ileum into the large pouch of the cæcum is protected by a *valve*, called the *ileo-cæcal*, which prevents regurgitation of its contents back into the small intestine; it is formed from the mucous coat of the cæcum, and, within that part of the intestine, appears as a transverse, elliptical opening, formed by two lips, which, when closed, overlap each other, and which any distension of the cæcum can only close more effectually.

The walls of the intestines are composed of four coats, viz: peritoneal, muscular, cellular, and mucous. The *peritoneal coat* completely surrounds the small intestine, its two layers meeting to form the mesentery. With the exception of the cæcum, which it wholly invests, and furnishes with a mesentery, it only partially covers the large intestine; passing over its anterior portion, and then being reflected on to the parietes of the abdomen, it leaves the posterior wall in direct contact with the iliac fascia. Attached to the peritoneal surface of the large intestine, small fatty bodies are sometimes seen, hanging off in a fringe-like manner; they vary in number and size, and are called *appendices epiploicæ*. If the peritoneum be peeled off from the intestine, the *muscular coat*, consisting of longitudinal and transverse fibres, will be seen beneath it; the longitudinal fibres are external to the circular, but the latter are thicker and most developed, except in the large intestine, where the muscular fibres are chiefly collected in the longitudinal bands characteristic of that part of the tube. The *mucous coat* is continuous throughout the whole alimentary canal. In the upper part of the small intestine it lies in a series of transverse folds, called *valvulæ conniventes;* these gradually disappear in the ileum. In the large intestine it is thrown into sharp ridges, corresponding to the constrictions of its sacculi. The sub-mucous cellular tissue connecting the mucous and muscular coats, which may be demonstrated by stripping off the mucous membrane, is sometimes described as a fourth coat, called the *cellular*.

The intestines are furnished with certain GLANDS, which may be seen upon the surface of the mucous membrane. The *solitary glands* are white, rounded, and slightly prominent bodies, scattered over the surface of both large and small intestine, in the former being found chiefly in the cæcum. The *follicles of Lieberkuhn*, found also in both large and small intestine, consist of little follicles with a minute orifice, hardly perceptible. The *agminated*, or *Peyer's glands*, belong chiefly to the ileum, and exist only in the small intestine; they are of an oblong shape, have a granulated aspect, and are placed opposite the attachment of the mesentery; the number and position of these patches vary, but there is almost always a large one near the termination of the ileum in the cæcum. The velvet-like surface of the mucous membrane of the small intestine is made up of *villi*, minute processes in which commence the *lacteals*, and which are only distinctly visible with the assistance of a lens.

The *duodenum*, not having been removed with the rest of the intestine, will be found in situ; it is the first ten or twelve inches of the canal beyond the stomach; it has no mesentery, and lies beneath the peritoneum, which only covers its anterior surface; the direction taken by it gives it the shape of a horseshoe, the convexity of which looks toward the right side; the head of the pancreas occupies the concavity, and its duct, as well as that of the liver, opens into it through the posterior wall. A form of glands, called *Brunner's glands*, is peculiar to this portion of the intestine; they are scattered in great numbers throughout its whole extent, and give the mucous surface a sort of granular aspect.

The STOMACH is an expanded portion of the alimentary canal, between the œsophagus and duodenum. The œsophagus, after passing through the diaphragm, terminates by gradually dilating into the stomach. That portion of the stomach expanding to the left of the œsophagus is called its *cardiac extremity*, and that at the opposite end the *pyloric extremity*. The superior curved border is called the *lesser curvature*, and the inferior border the *greater curvature;* to these borders are attached respectively the lesser and greater omentum. The stomach varies in size, according to the individual, and the degree of distension by its contents, and consequently varies somewhat in the position it occupies; it is, however, always in relation with the diaphragm, the abdominal parietes, and the liver, and by a fold of the peritoneum, called the *gastro-splenic omentum*, with the hilus of the spleen. The left pneumogastric nerve may be traced upon the anterior surface of the stomach, and the right upon its posterior surface.

The stomach is to be removed by dividing it at the œsophagus, close to the diaphragm, and at the duodenum, in such a way as to leave behind that portion perforated by the ductus choledochus communis; it should then be laid open along the greater curvature.

The coats of the stomach, as in the rest of the alimentary canal, are four in number, viz: peritoneal, muscular, cellular, and mucous. The peritoneal coat envelops it entirely. The muscular coat is made up of numerous and well-marked fibres, disposed in longitudinal, circular, and oblique directions. The mucous coat is movable, and slides upon the muscular, from which it is separated by the cellular coat; if not fully distended it will therefore be thrown into rugæ by the contraction of the latter; its surface is characterized by a sort of granulated appearance due to the follicles with which it is provided. At the pyloric extremity the mucous membrane forms a valve called the *pylorus;* this is sometimes a complete ring, at others merely two crescentic folds; the muscular fibres beneath form a sort of sphincter, and the entrance to the duodenum may be closed by their contraction. The color and condition of the mucous coat is variously modified by the contents of the stomach and by post-mortem changes.

THE SPLEEN.

The SPLEEN is situated beneath the cartilages of the ribs, at the cardiac extremity of the stomach, with which it is connected by means of the gastro-splenic omentum; it is also held in contact with the diaphragm by a suspensory peritoneal ligament; its connections, however, are not very strong, and it may easily be torn out from its position with the fingers. It is completely invested by peritoneum, which is closely adherent, and presents a smooth surface, convex externally, and concave internally. The upper end of the spleen is larger than the lower; the posterior border is blunt and rounded, the anterior comparatively sharp, and containing one or more notches. The concavity of the spleen admits the arteries, and gives exit to the veins, and that portion in relation with these is called its *hilus*. The splenic artery, of large size, divides at its entrance into several branches; the splenic vein, also of large size, is joined by the gastric, and superior, and inferior mesenteric veins, and terminates in the portal vein. Its nerves are derived from the solar plexus.

A section of the spleen exhibits a reticulated parenchyma of very elastic nature, from the meshes of which large quantities of blood may be squeezed or washed. In its tissue, pearly bodies are sometimes seen, not unlike miliary tubercles; these are called the *corpuscles of the spleen.*

The spleen varies in weight from two drachms to forty pounds; its average is from five to seven ounces. Its size may be averaged at five inches in length by three or four in width, and one to two inches in thickness. Within the folds of the peritoneum, in the neighborhood of this viscus, it is not uncommon to find *supplementary spleens*, small rounded bodies, varying from the size of a pea to that of a pigeon's egg; in structure these are similar to that of the spleen proper.

THE PANCREAS.

The PANCREAS is an irregularly shaped viscus, some six inches in length, lying transversely upon the vertebræ, behind the stomach. It is a conglomerate gland, the substance of which is of a pale color, and broken up into easily separated lobules. It is divided into a *head* and *body*, the head being that portion embraced by the duodenum. A portion of the gland surrounds the mesenteric vessels, and this prolongation is sometimes called the *lesser pancreas*. By making a longitudinal incision into the substance of the pancreas, and carefully separating the lobules, a delicate white duct will be exposed; this is the *pancreatic duct;* it commences at the extremity of the gland, and gradually increasing in size by the accession of small branches, makes its exit at the head of the viscus, and, in company with the ductus choledochus communis, penetrates obliquely through the walls of the duodenum at the posterior part of its perpendicular portion, opening into its interior by a common orifice with the bile duct. Occasionally it has a separate aperture, and there may be a second, supplementary duct.

THE LIVER.

By lifting the cartilages of the right side with the chain hooks, and by lifting and drawing the liver in different directions, its attachments and relations may be successively observed and studied.

The LIVER is a conglomerate gland, and the largest viscus in the body; it weighs from three to four and a-half pounds; its color is a reddish-brown, and its surface is covered with peritoneum, except in such intervals as are left between the layers reflected from the viscus to the parietes of the abdomen, and which constitute the ligaments which hold it in place. It is in relation superiorly and anteriorly with the diaphragm and abdominal parietes; inferiorly with the stomach and transverse colon, and posteriorly with the diaphragm and vertebral column, the aorta, and vena cava.

The liver is held in its place by the following ligaments:—

| Broad, or Suspensory, | Right and left Lateral, |
| Round, | Coronary. |

The *broad*, or *suspensory ligament*, with its accompanying round ligament, was noticed at the time the abdomen was opened, and their divided ends will now be seen lying upon the superior surface of the liver at its anterior margin. It is a thin, double layer of peritoneum, extending from the posterior to the anterior border of the liver, and attached to the diaphragm and anterior wall of the abdomen, as far as the umbilicus.

The *round ligament* is a white fibrous cord, the obliterated umbilical vein of fœtal life, occupying the lower border of the broad ligament; it passes under the anterior border of the liver into a longitudinal fissure of its inferior surface, terminating in the walls of the portal vein.

The *lateral ligaments* are peritoneal folds, passing from each extremity of the liver to the diaphragm, the left being the longest of the two.

The *coronary ligament* is situated along the posterior border of the viscus, between the two lateral ligaments. The two reflections of the peritoneum composing it leave a considerable space between them, so that this portion of the liver is in immediate contact with the diaphragm, and connected with it by firm cellular attachments.

To remove the liver, these different ligaments are to be divided; it is then to be carefully dissected from the diaphragm, cutting across the vena cava where it perforates the muscle, and preserving a portion of it in connection with the liver. Precaution should be taken not to tear the liver or perforate the diaphragm, accidents which are liable to occur. The duodenum, or that portion of it left expressly on account of its connections with the liver, should also be removed together with the ducts which connect it to the gall-bladder and inferior surface of the liver.

Removed from the body, the liver presents superiorly a surface uniformly convex, and divided into two portions by the broad ligament, called *right* and *left lobes*, the right being much the larger of the two. It can now be seen that the posterior border is rounded, and the anterior sharp and thin; the posterior border has a depression where it rests upon the vertebral column; the anterior border is notched at the point of its separation into two lobes, and sometimes also in front of the gall-bladder.

The inferior surface of the liver is very irregularly concave, and is broken up by fissures and lobes. A fissure containing the round ligament divides the under surface, as the broad ligament does the upper, into its right and left lobes; the left lobe is comparatively thin and smooth, the right is much thicker; upon it are situated three other lobes, viz.: the lobus Spigelii, caudatus, and quadratus.

The *lobus Spigelii* is a triangular-shaped, partially detached lobe, surrounded by fissures, and placed in the middle of the posterior part of the liver.

The *lobus caudatus* is a sort of ridge, extending from the lobus Spigelii toward the middle of the right lobe; it is not usually well defined.

The *lobus quadratus* is the portion of the right lobe intervening between the gall-bladder and the fissure of the round ligament, and in front of the fissure at which enter the portal vein and hepatic artery.

The *transverse fissure* is the most important of all the fissures of the under surface of the liver; it occupies nearly the centre of the organ, and is the depression where the portal vein and hepatic artery enter, and the hepatic duct makes its exit. It is sometimes called the *hilus*, and as being the entrance or *porta* of the liver, gives a name to the vein which finds admission at this point. The hepatic duct is the most anterior of the vessels of the transverse fissure, the artery is placed next, while the vena portæ is the most posterior.

The *longitudinal fissure*, containing the round ligament, joins the transverse nearly at a right angle; the substance of the liver sometimes crosses this fissure, and converts it partially into a canal.

The *fissure* of *the ductus venosus* is that portion of the longitudinal fissure posterior to the transverse; it lies between the lobus Spigelii and the left lobe, and contains a rounded cord, continuous with the round ligament, which terminates in the vena cava; this cord being the obliterated ductus venosus, which, during fœtal life, conveys directly to the vena cava a portion of the blood of the umbilical vein.

The *fissure* of *the vena cava* lies between the lobus Spigelii and the right lobe; it is occupied by the vena cava, and is sometimes converted into a canal. On laying open the vena cava, the orifices of the *hepatic veins* will be seen, this being the point at which they discharge their blood into the general venous circulation.

The *fissure* of *the gall-bladder* is the fossa in which that receptacle rests.

From the preceding description, it will be seen that the liver has five lobes, five fissures, five ligaments, and five vessels, which may thus be tabulated:—

Lobes.	*Fissures.*	*Vessels.*	*Ligaments.*
Right,	Transverse,	Hepatic Duct,	Coronary,
Left,	Longitudinal,	Hepatic Artery,	Broad,
Spigelii,	Ductus Venosus,	Hepatic Veins,	Right Lateral,
Caudatus,	Vena Cava,	Portal Vein,	Left Lateral,
Quadratus.	Gall-Bladder.	Vena Cava.	Round.

The GALL-BLADDER is a pyriform sac, the fundus of which extends just beyond the anterior margin of the liver; the neck is directed toward the transverse fissure, and terminates in the cystic duct. The coats of the gall-bladder are three in number, viz., peritoneal, cellular, and mucous. The *peritoneal coat* only partially invests it, the posterior surface being in direct contact with the liver, and connected to it by the *cellular coat.* The *mucous coat* is very finely reticulated, and stained deep brown by the bile which tinges all the coats, and gives the bladder a greenish hue; toward the neck, the mucous membrane is thrown into prominent folds, which form a sort of valve at the point of its union with the cystic duct.

The *cystic duct*, about an inch in length, joins the *hepatic duct* of the liver two inches distant from the transverse fissure at which it emerges; their union forms the *ductus choledochus communis*, a tube of some three inches in length, which terminates in the duodenum as already described. The gall-bladder is merely a reservoir of superfluous bile, and the ductus choledochus conveys the bile from it as well as that derived directly from the liver.

The hepatic duct, the hepatic artery, and portal vein, are surrounded by a loose areolar tissue which accompanies them in their ramifications, and which is called *Glisson's capsule.* This capsule may be seen, constituting, as it were, the fibrous skeleton of the organ, by tearing the liver, or by stripping off the closely adherent peritoneal coat; it surrounds the minute granules or *acini*, of which this fracture shows the liver to be made up.

The hepatic veins are closely adherent by their parietes to the substance of the liver, and cannot, therefore, contract, as the loose areolar tissue surrounding the portal veins permits those vessels to do; in a section of the liver, consequently, these may be easily distinguished from each other. The portal vein will be found collapsed, and accompanied by an artery and a duct; while the hepatic vein will remain open and uncontracted, and is unaccompanied by any other vessel.

The liver is supplied with blood from the hepatic branch of the cœliac axis; the portal vein also conveys its current through the organ, and, collecting the impure blood of the hepatic arteries, terminates in the hepatic vein; this empties into the vena cava inferior, and so completes the hepatic circulation. The *hepatic plexus* of nerves, derived from the solar plexus, accompanies the hepatic artery in its subdivisions. The liver also receives branches from the phrenic and pneumogastric nerves.

DISSECTION IV.

THE SUPRA-RENAL CAPSULES.

The supra-renal capsules are to be sought for beneath the peritoneum, high up in the lumbar region, on each side of the spinal column.

The SUPRA-RENAL CAPSULES are small, flat, crescentic bodies, covered in by peritoneum and surmounting the kidneys, to which their concave surface is sometimes directly applied; an interval not unfrequently, however, exists between the two organs. They are of a yellowish color, and easily obscured among the fat and cellular tissue of the surrounding parts. The right capsule rests upon the diaphragm between the kidney and the liver, oftentimes as closely adherent to one as the other of these organs. The left capsule also rests upon the diaphragm, beneath the pancreas and spleen. For so small organs they are largely supplied both with nerves and arteries, the former constituting a *supra-renal plexus* derived from the solar plexus, the latter being branches of the aorta, and the phrenic and renal arteries; the supra-renal vein terminates sometimes in the inferior cava and sometimes in the renal vein. Upon section, they are seen to be made up of an external or cortical substance, of a yellow color, and an internal or medullary substance, dark and semi-fluid; an internal cavity is sometimes described, but it is a question if it be not the result of post-mortem degeneration. One or two instances of entire absence of the supra-renal capsules have been reported, and in the fœtus they have been found fused into one body across the vertebral column.

THE KIDNEYS.

The kidneys occupy the lumbar region on each side of the vertebral column, lying beneath the peritoneum, enveloped in fat and cellular tissue. To expose them, both the peritoneum and lumbar fascia must be removed, and the ureter should be traced downward, either in part or the whole of its course.

The KIDNEYS are oval-shaped organs of a deep red color, and smooth surface, convex externally and concave toward the vertebral column; the concave border at its central part presents a longitudinal fissure called the *hilus;* at this point the renal artery enters and the vein and ureter emerge.

The URETER is the excretory duct which conveys the urine to the bladder; it is of the size of a pipe-stem, except at its commencement, where it is dilated and forms part of what is called the *pelvis* of the kidney. It is of a bluish-white color and of a firm fibrous texture, and is lined within by a thin mucous membrane; it passes downward upon the psoas muscle, behind the peritoneum, crosses the external or common iliac artery, and dips down behind the bladder to enter it at its base. The ureter is sometimes double, either in a part or the whole of its length.

The ureter should be divided midway between the kidney and the bladder; then the vein and artery being cut across, the kidney may be easily, and without injury, torn from its bed by the fingers. A longitudinal incision along its convex margin should divide the kidney in such a way that it may be opened and its interior exposed, without its two halves being separated at the concave border.

The kidney is about four inches long, two wide, and one in thickness; it weighs about four and a half ounces. Its upper extremity is larger than the lower; its posterior surface is flatter than its anterior, and it is invested with a special capsule, which the nail of the thumb and fore-finger may easily detach from its surface. Internally, it is composed of two distinct structures and an irregular-shaped cavity, called the pelvis. That portion surrounding the pelvis is made up of cones, six or eight in number, called *pyramids of Malpighi*, the apices of which are directed toward the centre of the organ; they are composed of a congeries of straight tubes called *tubuli uriniferi*. The apices of the pyramids are called *papillæ*, and are covered by a thin mucous membrane, perforated by orifices, through which the tubuli discharge the urinary secretion into the pelvis; they are surrounded by a sort of "prepuce," so to speak, also covered by a thin mucous membrane, and which is called the *calyx*, or *infundibulum*, of the pyramid; the projection of the papillæ into the pelvis produces the irregularity of outline which characterizes that cavity; the ureter opens from the general cavity of the pelvis, and their mucous surfaces are continuous. The external, or cortical portion of the kidney, surrounds the bases of the pyramids, and penetrates irregularly between them under the name of the *columns of Bertin;* it is composed of the tubuli convoluted among minute vascular ramifications.

The kidney is a very vascular organ and receives its blood through the renal artery, a large branch of the aorta, affording an excellent illustration of the law that the size of an artery is in proportion, not to the size of the viscus to which it is distributed, but to the activity of its functions. The renal vein empties into the inferior vena cava; and the nerves, which are abundant, are received from the renal plexus, this being derived from the solar plexus and the lesser splanchnic nerve. The large upper extremity, the flattened posterior surface, together with the relative position of the vessels at the hilus permit the right kidney to be distinguished from the left after their removal from the body; the renal vein is the most anterior, the artery is behind the vein, and the ureter posterior to them both.

The kidneys not unfrequently present deviations from their normal shape and position; sometimes they are united into one body across the vertebral column, forming what is called a *horse-shoe kidney;* occasionally one is wanting. They may be found placed much lower in the lumbar region than usual, and have even been found in the pelvis. They are usually firmly fixed in their positions, but in certain instances a degree of mobility has been noticed, appreciable by palpation during life, and constituting what is called a *floating kidney.*

THE DIAPHRAGM.

To get a good view of the diaphragm, a large block is to be placed under the loins in such a way as to elevate the base of the thorax. The peritoneum must be thoroughly removed from its surface; and if the thorax has not yet been opened, it can be readily stripped off by the forceps and fingers.

The DIAPHRAGM is a muscular plane separating the abdomen from the thorax; it arises from the sternum by short and separated fibres, which leave a triangular interval between them, covered only by the peritoneum on one side and the pleura on the other; from the superior border and internal surface of the last six ribs by serrations which

17

indigitate with the transversalis muscle of the abdomen (p. 166); also from a fibrous arch called the *ligamentum arcuatum externum*, extending from the tip of the last rib to the transverse process of the first lumbar vertebra, and curving over the quadratus lumborum muscle; it further arises from a smaller fibrous arch called the *ligamentum arcuatum internum*, which stretches from the termination of the preceding to the body of the second lumbar vertebra, curving over the psoas magnus muscle; and also from the bodies of the second and third lumbar vertebræ by tendinous fibres which are common to it and the anterior common ligament of the vertebræ. From this circumference the fibres converge to be inserted into the *central tendon*, a glistening expansion in the centre of the muscle, the metallic brilliancy of which has given it the name of *speculum of Van Helmont*. The above description is of what is called the *larger muscle* of the diaphragm. The *lesser muscle* consists of two fleshy bundles arising by separate tendons from the lumbar vertebræ; the right is the larger and the longer; it is connected with the bodies and intervertebral cartilages of the upper four lumbar vertebræ; the left bundle takes its origin from only the upper three; both are inserted into the central tendon. These tendons are also called the *pillars* or *crura* of the diaphragm, and the left one is sometimes wanting. The right pillar crosses the left pillar in such a way as to divide the interval between them into two separate parts; the lower, parabolic in shape, gives passage to the aorta, vena azygos, and thoracic duct; the upper, elliptical in shape, transmits the œsophagus and the pneumogastric nerves. In the central tendon is a third opening, the largest of all, which transmits the vena cava inferior.

The diaphragm, on the thoracic side, is covered by the pleura and the pericardium, the fibrous layer of which blends with the central tendon.

The diaphragm is supplied by the phrenic nerves, and is nourished by the phrenic arteries and by the musculophrenic branch of the internal mammary artery.

The Receptaculum Chyli, ordinarily the commencement of the thoracic duct (p. 123), lies close beside the right pillar of the diaphragm, and between the aorta and vena cava; it usually rests upon the second lumbar vertebra, and consists of a thin, semi-transparent, membranous sac, oblong and irregularly shaped; it is usually empty, and may be best demonstrated by insufflation with the blowpipe from the duct above. The thoracic duct may sometimes, though rarely, be traced below the receptaculum, which, in that case, becomes merely a dilated portion of the duct.

The *greater and lesser vena azygos* (p. 123) may sometimes be better seen in this connection than at any other period of the dissection. Both these veins usually pass through the aortic opening of the diaphragm, but they not unfrequently perforate its fibres at the side of the pillars.

SUPERFICIAL FEMORAL REGION.

The knees of both legs should be bent so as to bring the soles of the feet in apposition, with the heels approached to the nates; an incision, six inches in length, is to be made through the skin from the centre of Poupart's ligament down the thigh, and the integument reflected to either side; this will expose the superficial fascia.

The *superficial fascia* of the thigh, like that of the abdomen, with which it is continuous, consists of two layers; it contains several small arterial branches, all arising from the femoral artery just below Poupart's ligament.

The *superficial epigastric artery* passes upward, between the two layers of the fascia, toward the umbilicus, and has already been seen, in part, during the dissection of inguinal hernia.

The *superficial external circumflex artery* passes outward toward the anterior superior spinous process and crest of the ilium.

The *superficial external pudic artery* passes inward to be distributed to the integument of the penis and scrotum, or to the labia majora in the female.

Several *cutaneous nerves* will be found lying in the superficial fascia; they are small, and derived from the anterior crural and genito-crural branches of the lumbar plexus.

Enlarged lymphatic glands are usually found both in and beneath the superficial fascia; they are connected with the lymphatics of the lower extremity, and their efferent ducts enter the abdomen through the saphenous opening.

The superficial fascia may be removed by dissecting it away from the fascia lata beneath; a portion of it should, however, be left for the present, around the point where the saphena vein penetrates the thigh.

The *deep external pudic artery*, arising from the femoral, lies between the superficial and the deep fascia; after a short course toward the inside of the thigh it penetrates the deep fascia, and then, reappearing from beneath it, is distributed to the external organs of generation. It is the largest of the superficial arteries.

ANATOMY OF FEMORAL HERNIA.

Inasmuch as the dissection of the internal abdominal muscles, and the parts in relation to them, will destroy many of those connected with the anatomy of femoral hernia, it is desirable to proceed to its examination before they are disturbed. Femoral hernia is most advantageously studied upon a female subject.

The INTERNAL SAPHENA VEIN, commencing on the back of the foot, ascends along the inside of the thigh to join the femoral vein; that part of it exposed in the present dissection, lies between the superficial and the deep fascia, and receives several small veins from the neighborhood; it passes through an opening in the deep fascia, called the saphenous opening, and then unites with the femoral vein. It is at this opening that *femoral*, or, as it is sometimes called, *crural hernia*, takes place. A portion of the superficial fascia, perforated by small orifices, which give entrance to the efferent ducts of the lymphatic glands imbedded in it, and hence called the *cribriform fascia*, covers in the saphenous opening.

The *deep fascia*, or *fascia lata*, is a strong fibrous sheath investing the whole lower extremity. The *saphenous opening*, a little to the inside of the middle of the thigh, and an inch below Poupart's ligament, ovoid in shape, and in outline resembling the reversed Greek letter sigma, divides the fascia lata into two parts, iliac and pubic. The *iliac* portion is on the outer side, and is connected with the whole length of Poupart's ligament; from this attachment it passes outward and downward, its edge forming the outer margin of the saphenous opening, and constituting what is called its *falciform border*. This border may be followed round, underneath the saphena vein, and will be found to unite with the pubic portion of the fascia lata on the inside of the opening; it also blends with the superficial fascia, so that they are only separable by the knife. The *pubic* portion of the fascia lata is inserted into the spine of the os pubis, and into the pectineal line, where it becomes continuous with the fascia investing the psoas and iliacus muscles within the abdomen. Poupart's ligament, attached to the spine of the os pubis, also extends inward three-fourths of an inch, forming a triangular expansion inserted into the pectineal line and having a concave border directed toward the femoral vessels. This triangular portion is known as *Gimbernat's ligament*. A portion of the fascia lata blends with this under the name of *Hey's ligament*.

Reverting to the abdominal cavity, and removing the peritoneum, it will be seen that the transversalis fascia (p. 169), and the fascia covering the psoas and iliacus muscles unite, externally to the femoral vessels, and form a continuous fold, closely connected with Poupart's ligament; where the vessels pass out these fasciæ separate and surround them, and blend with their areolar sheath about two inches below the ligament; the sort of funnel thus made, and through which the vessels pass downward into the thigh, is called the *infundibuliform fascia*. It is plain, therefore, that while the fold protects the *crural arch*, which is the space between Poupart's ligament and the os innominatum, from a hernial protrusion externally, it is liable to occur internally where the exit of the vessels obliges these fasciæ to separate.

The crural arch is occupied, externally, by the psoas and iliacus muscles, between which passes the crural nerve; next the muscles comes the external iliac artery, then the external iliac vein, between which and the inner termination of Poupart's ligament is a space called the *crural ring*. The crural ring is about half an inch wide, and is filled with the sub-peritoneal cellular tissue; that portion of this tissue which stretches over the crural ring, and in which there is often found a lymphatic gland, is called the *septum crurale*. The greater size of the crural arch in females than in males, owing to the greater breadth of their pelvis and the lesser development of the soft parts, makes the crural ring a much larger space in them, and explains why they are more subject than males to hernia at this point. The vessels and the crural ring are all inclosed in the infundibuliform fascia, but are separated from each other by septa. By dividing Poupart's ligament, and turning it carefully to either side, these septa may be seen, one between the artery and vein, and one between the vein and the crural ring. The space between the crural ring and the point at which the saphena vein joins the femoral, is called the *crural canal;* its anterior wall is the transversalis portion of the infundibuliform fascia, with the falciform border of the iliac side of the fascia lata; the posterior wall is formed by the iliac portion of the infundibuliform fascia and part of the pubic side of the fascia lata; its external wall is the septum, forming the division between the canal and the vein, and its internal wall, Gimbernat's and Hey's ligament, covered with their portion of the infundibuliform fascia. The epigastric artery is in close relation to the upper and outer side of the crural ring, and the obturator artery, when given off from the external iliac, dips downward along its outer and inferior edge; in a certain number of instances this artery arises from a trunk common to it and the epigastric, in which case it curves around the ring to descend upon its inner side; the obturator vein may follow the same course. It will be seen that this distribution of the artery exposes it to be wounded in dividing the crural ring, in an operation for strangulated femoral hernia; observation, however, having shown that the point of constriction is ordinarly in the fascia lata and not in the ring, the danger, which formerly was so properly dreaded, has disappeared before a more enlightened practice.

The spermatic cord is separated from the crural ring by Poupart's ligament.

The preceding description shows that a knuckle of intestine, entering the crural ring, finds little or nothing to oppose its protrusion at the saphenous opening, externally; and although while in the canal it must be much constricted, from the unyielding nature of its walls, the hernia may, after emerging, expand into a tumor of considerable size. The blending of the superficial fascia with the fascia lata, just below the saphenous opening, prevents the hernia from descending below that point, and arrived at this obstruction, it turns upward toward Poupart's ligament. The practical importance of this fact, in connection with attempts at reduction by taxis, is of course obvious; to return the hernia within the abdominal cavity, pressure must first be exerted downward, then backward and upward.

The intestine, in descending, must carry before it the following coverings; first, the peritoneum, then the septum crurale, and that portion of the infundibuliform fascia which, forming the anterior wall of the canal, is perforated by the saphena vein, these two latter constituting what is called the *fascia propria* of the hernia; and, externally to the saphenous opening, the cribriform fascia, the superficial fascia, and the integument.

THE LUMBAR PLEXUS.

The dissection of the lumbar plexus presupposes the removal of the peritoneum and cellular tissue in the lumbar region; the nerves composing it are mostly small in size, and though liable to be divided in the dissections already made, their origin and a portion of their course may always be demonstrated.

The LUMBAR PLEXUS is formed by the communications of the anterior branches of the five lumbar nerves; the posterior branches being distributed to the muscles of the back. The principal trunks are the following, viz:—

Musculo-cutaneous,	Crural,
External Cutaneous,	Obturator,
Genito-crural,	Lumbo-sacral.

The *musculo-cutaneous nerve*, coming from the first lumbar nerve, crosses the quadratus lumborum muscle obliquely, to reach the middle of the crest of the ilium, where it pierces the transversalis muscle, and divides into two branches, the abdominal and the scrotal; the *abdominal* supplying the muscles and integument of the abdomen; the *scrotal*, joining the spermatic cord in the male, and the round ligament in the female, at the external inguinal ring, is distributed to the integument of the scrotum, or, in the female, to the labia majora.

The *external cutaneous nerve*, coming from the second lumbar nerve, crosses the iliacus internus muscle obliquely, to reach the anterior superior spinous process of the ilium, where it passes underneath Poupart's ligament, and is distributed to the integument of the gluteal region, and the outside of the thigh.

The *genito-crural nerve*, coming from the second and third lumbar nerves, runs down upon the psoas muscle, and divides into a genital and crural branch. The *genital* branch enters the internal abdominal ring, and accompanies the spermatic cord, or round ligament, to the integument of the groin; the *crural* descends along the outer border of the external iliac artery, enters the sheath of the femoral vessels, and is distributed to the integument of the front of the thigh.

The *crural nerve*, coming from the second, third, and fourth lumbar nerves, is the largest branch of the lumbar plexus; it pierces the psoas muscle, then passes downward between it and the iliacus internus muscle, and, about an inch below Poupart's ligament, divides into numerous branches hereafter to be described. (p. 223.)

The *obturator nerve*, coming from the third and fourth lumbar nerves, passes down among the fibres of the psoas muscle, and behind the iliac vessels, along the brim of the pelvis to the upper and inner part of the obturator foramen, which it perforates, to be distributed to the adductor muscles of the thigh. An *accessory obturator* branch sometimes arises by separate filaments from the third and fourth lumbar nerves, or from the upper part of the obturator nerve itself; it passes down along the inner border of the psoas muscle, and is interesting only because it supplies the hip-joint, which it enters beneath the transverse ligament, in company with the articular branch of the internal circumflex artery. When the accessory nerve is wanting, the hip-joint is supplied by a branch of the obturator nerve.

The *lumbo-sacral nerve*, from the fourth and fifth lumbar nerves, descends into the pelvis over the base of the sacrum, and unites the lumbar with the sacral plexus.

The dissection of these nerves will have exposed the internal abdominal muscles.

The QUADRATUS LUMBORUM is covered in by the anterior layer of the lumbar fascia; this being removed, it will be found to arise from the last rib, and the transverse processes of the upper four lumbar vertebræ, and to be inserted into the posterior part of the crest of the ilium. If the muscle is divided, and the two ends reflected, the middle portion of the lumbar fascia will be exposed.

The PSOAS PARVUS MUSCLE lies in front of and upon the psoas magnus; it arises from the sides of the bodies of the last dorsal and first lumbar vertebræ, forming a small belly which expands into a broad, flat tendon, which, losing itself in the fascia of the iliacus muscle, is inserted into the pectineal line and ilio-pectineal eminence in a manner calculated to prevent the contractions of the psoas magnus muscle from compressing the iliac vessels. This muscle is frequently wanting; when absent the fascia iliaca is more developed, and supplies its place.

The PSOAS MAGNUS MUSCLE is a long muscle, lying parallel to the vertebral column, and arising from the sides of the bodies of the last dorsal, and four upper lumbar vertebræ, and from the transverse processes of all the lumbar vertebræ; it forms the border of the true pelvis laterally, and is inserted, in common with the iliacus internus muscle, into the trochanter minor of the femur, and an inch or more of the shaft of the bone below it. That portion arising from the transverse processes is sometimes distinct from the rest of the muscle in its whole course.

The ILIACUS INTERNUS MUSCLE arises from the transverse process of the last lumbar vertebra, the crest and concavity of the ilium, and the anterior part of the capsule of the hip-joint, and in common with the tendon of the psoas magnus, is inserted into the lesser trochanter of the femur. The insertions of neither of these muscles can be satisfactorily seen until after the dissection of the muscles of the anterior femoral region.

The EXTERNAL ILIAC ARTERY, with its accompanying vein, lies along the inner border of the psoas muscle; it has no branches till it reaches Poupart's ligament, where it gives off two, and sometimes three. About half an inch outside the artery as it passes under Poupart's ligament, the crural nerve may be seen, lying deep between the psoas and iliacus muscles.

The *epigastric artery* has been already noticed (p. 167), but may now be more particularly observed as to its origin from the external iliac; this is usually close to Poupart's ligament, but it may be found at a very considerable distance either above or below it; it frequently furnishes the obturator artery.

The *circumflexa ilii artery* arises from the outer side of the artery, and winds along Poupart's ligament, and the crest of the ilium; it lies between the internal oblique and transversalis muscles, and breaks up into numerous branches, some of which supply the muscles, while others inosculate with the ilio-lumbar artery, a branch of the internal iliac; these anastomoses principally take place beneath the iliacus muscle, and if that be divided and detached from the ilium, they will be brought into view.

The *obturator artery* nominally arises from the internal iliac, but in a certain number of cases it arises from the external iliac by an independent origin, or by a trunk common to it and the epigastric; it then descends to reach the upper and inner part of the obturator foramen, and passes out through an opening in the obturator membrane, to be distributed outside the pelvis: as it perforates this membrane it is often joined by a small branch given off from the internal iliac, which preserves the normal origin and course of the artery.

It will be desirable at this period to turn the subject over. If the previous dissections have been accomplished, this can be done without embarrassing those engaged upon the lower extremity. The parts connected with the region of the back will be found described in Part Second, Dissection V.

DISSECTION V.

ANATOMY OF THE PERINEUM.

As the dissection of the perineum interferes with that of other parts of the body, it should consequently be made in common, the dissectors mutually agreeing to suspend operations until it is accomplished. It can be done advantageously only on the male subject; the peculiarities of the female perineum will be found at p. 205.

The legs being flexed, the thighs are to be bent upon the trunk, and the nates made to project over the edge of the table, preserving their position by one or two turns of a cord carried round the right knee, then under the table to the left knee, and finally made fast by again attaching it to the right knee. The subject being thus placed, the scrotum and testicles should be lifted on to the pubes, and kept out of the way by hooks or pins; the rectum is to be distended with cotton wool, tow, or similar material, and, when well filled, the anus should be made to project by pressing it downward from within the pelvis. The perineum being washed and shaved, is then ready for dissection. An elliptical incision, commencing at the root of the scrotum, its long diameter corresponding to the median line, should include the anus, and extend to the coccyx posteriorly; the integument is then to be dissected toward the anus.

The PERINEUM is an important surgical region, bounded on each side by the tuberosities of the ischia, in front by the arch of the pubes, and behind by the coccyx; in other words, its boundaries are those of the inferior strait of the pelvis.

Immediately beneath the skin, where it becomes continuous with the mucous surface of the anus, the fibres of the sphincter ani will present themselves, pale in color, and indistinctly characterized. Elsewhere, a layer of fat and cellular tissue, constituting the *superficial fascia*, covers in the deeper parts of the perineum; laterally, between the anus and the ischia this is considerable in amount, and fills a large space, called the *ischio-rectal fossa*.

The superficial fascia, consisting of two layers, is connected laterally with the rami of the ischia and the pubes, but through the medium of the scrotum becomes continuous with the corresponding structure in the groin (pp. 168 and 195). Posteriorly, it dips down in front of the rectum to join the anterior layer of the triangular ligament, or deep fascia of the perineum. Hence, abscesses of this region, or extravasation of urine from rupture of the urethra, do not extend backward behind the rectum, or laterally upon the thighs, but forward toward and into the scrotum, and even to the anterior part of the abdomen. In and beneath this fascia will be found a number of vessels and nerves.

The *external hemorrhoidal artery* traverses the ischiorectal fossa; it is an offset from the internal pudic artery, a branch of the internal iliac, which lies under the ramus of the ischium, and is distributed to the sphincter and levator ani muscles and to the lower part of the rectum. Farther in front, the internal pudic artery gives off the *superficial perineal artery;* this passes forward to the scrotum, giving off in its course a branch called the *transversalis perinei,* which crosses the perineum upon the transversus perinei muscle.

These arteries are accompanied by small branches from the internal pudic and perineal cutaneous nerves.

The *internal pudic nerve* is an offset from the sacral plexus, which takes the course of the internal pudic artery, and divides into a superior and inferior branch, the former going to the penis, the latter to the scrotum and perineum.

The *perineal cutaneous nerve* comes from the lesser sciatic, and ascends along the ramus of the ischium to supply the scrotum and the integument below the penis.

The preceding dissection will have exposed several muscles, small in size, but, with a single exception, sufficiently well marked to be readily recognized.

The SPHINCTER ANI is an elliptical-shaped muscle, surrounding the anus. It arises from the tip of the coccyx, and is inserted an inch or more in front of the anus in common with two other muscles, hereafter to be described, into a fibrous spot called the *perineal centre.* Deeper in the pelvis, but continuous with this muscle, will be seen the circular fibres proper to the rectum itself; these constitute the *sphincter ani internus;* by removing all the fat and cellular tissue from the ischio-rectal fossa, it will be seen that these muscles blend with, and become lost in the lower part of the levator ani muscle, which comes down from each side of the pelvis, and surrounds the gut, as will hereafter be observed on making a section of the pelvic cavity.

The TRANSVERSUS PERINEI MUSCLE is a small bundle of fibres, occasionally wanting on one or both sides, which arises from the tuberosity of the ischium, and is inserted into the perineal centre. A slip, arising in common with this, sometimes passes forward, and becomes blended with the accelerator urinæ; this is called the *transversus perinei alter.*

The ERECTOR PENIS arises from the tuberosity and ramus of the ischium by a strong tendon, and forming a round, fleshy belly, is inserted on the side of the penis into the strong fascia investing the corpus cavernosum.

The ACCELERATORES URINÆ lie upon the corpus spongiosum of the penis; arising from the perineal centre and the raphé that separates them, their fibres diverge to encircle the penis; the posterior fibres are inserted into the ramus of the pubes and ischium, the middle surround the corpus spongiosum, and the anterior, spreading upon the corpora cavernosa, are inserted into their investing fascia.

By dividing the muscles inserted into the perineal centre, and reflecting them, the bulb of the corpus spongiosum penis will be exposed, and, directly behind it, a strong fascia named the triangular ligament, through which the membranous portion of the urethra passes.

The TRIANGULAR LIGAMENT, so called from its occupying the triangular space formed by the arch of the pubes, is the deep perineal fascia, extending from one ramus of the ischium and os pubis to the other; it is composed of two layers, the anterior of which unites with the superficial fascia in front of the anus, and the posterior, passing backward, invests the membranous urethra and prostate, and becomes continuous with the pelvic fascia. It furnishes one of the chief supports and means of resistance to the superincumbent weight of viscera pressing down upon the perineum. The anterior layer of the triangular ligament is traversed by the urethra, and the edges of the opening through which it passes are continuous with the fibrous sheath of the corpus spongiosum. Between the two layers lie the compressor muscle of the urethra, Cowper's glands, and, for a part of their course, the external pudic arteries, and arteries of the bulb.

The *artery of the bulb* is a branch of considerable size, arising from the internal pudic; it passes transversely inward to the bulbous portion of the corpus spongiosum, to which it is distributed.

COWPER'S GLANDS, one on each side, are small bodies, the size of a pea, situated behind the bulb of the corpus spongiosum, between the two layers of the triangular ligament; they secrete a fluid carried by a duct into the bulbous portion of the urethra. Their size, and the nature of the locality in which they exist, often render them difficult to demonstrate.

The erector penis muscle is to be dissected away on one side, from the ramus of the pubes and ischium, keeping close to the bone; this will permit the dissection of the internal pudic artery.

The removal of the erector penis shows the strong tendinous nature of that muscle at its origin from the bone;

the other extremity being connected with the corpus cavernosum, it forms one of the chief supports of the penis, and is called the *crus penis*.

The INTERNAL PUDIC ARTERY may now be sought for, and will be found lying under the edge of the ramus of the ischium. This artery is one of the terminal branches of the internal iliac; it passes out at the greater sacro-ischiatic foramen, crosses the spine of the ischium, and enters the pelvis again by the lesser sacro-ischiatic foramen to reach the ramus of the ischium, about an inch in front of the tuberosity; it passes along beneath the edge of this to the symphysis pubes, where, under the name of the *dorsalis penis*, it runs the length of the organ from which it takes its name, to terminate in the glans penis.

The removal of a calculus from the male urinary bladder is accomplished by an operation performed in the perineum, called lithotomy. The bladder is reached by an incision extending from the median line, an inch or more in front of the anus, backward and outward, to a point midway between the anus and the tuber ischii. A part of the accelerator urinæ near the bulb of the corpus spongiosum, the transversus perinei muscle, and the transverse and superficial perineal arteries, besides the skin, superficial fascia, and triangular ligament, are necessarily divided, in order to reach the membranous urethra. The knife of the surgeon is directed by a sound introduced per urethram, and by the guidance of this the membranous urethra and prostate are incised, so that an opening for the extraction of the stone is effected. The artery of the bulb may be wounded in this operation, but the internal pudic artery can never be touched unless the incision is carried too near the ramus of the ischium.

The muscles of the female perineum differ little, except in name, from those of the male.

The *constrictor vaginæ* surrounds the orifice of the vagina, arising from the perineal centre, and is inserted into the corpus cavernosum of the clitoris; it corresponds to the accelerator urinæ.

The *transversus perinei* is inserted into the side of the constrictor vaginæ.

The *erector clitoridis* arises from the ramus of the ischium, and is inserted into the side of the corpus cavernosum of the clitoris.

The *artery of the bulb* is distributed to the vagina.

DISSECTION VI.

INTERIOR OF THE PELVIS.

To examine the interior of the pelvis it is necessary to detach one of the inferior extremities with the corresponding os innominatum. To do this, the symphysis pubes is to be cut through, leaving the penis and scrotum attached to the side to be preserved; the sacro-iliac articulation being found, and its anterior ligaments divided, by making the edge of the table a fulcrum and forcibly separating the divided pubes, the dislocation of the ilium from the sacrum will be effected; the common iliac and the branches of the internal iliac are to be divided, and the pelvic viscera separated from their lateral attachments on one side only; then cutting through the sacro-ischiatic ligaments and the gluteus maximus muscle and skin externally, the limb will be separated and one-half of the pelvis remain undisturbed. The disadvantage of this method is that one side is necessarily sacrificed. The sacrum may be sawed through upon the median line instead of dislocating the ilium; the glutei muscles are then left uninjured upon both sides, but the pelvis is not in so favorable a condition for advantageous dissection.

The position of the pelvic viscera, and the folds of peritoneum which invest them may now be studied.

It will be seen that the rectum is covered with peritoneum, and held in its place along the middle of the sacrum by a mesentery called the meso-rectum; from the rectum the peritoneum passes over the bladder, leaving a fold between called the *recto-vesical fold*, and which sometimes forms a tight band or cord-like edge on its posterior surface. If the subject be a female one, we shall have the uterus between the bladder and rectum, and there will then be two folds, the *recto-uterine* and the *vesico-uterine*. The lateral reflections of the peritoneum form the false ligaments which sustain the pelvic viscera. It will be noticed that the lower part of the rectum and a large part of the lower half of the bladder have no peritoneal coat, that membrane merely covering them in, and then being reflected to the sides of the pelvis. The point at which it is reflected from the bladder to the anterior abdominal parietes should be especially examined, to notice the fact that it is possible to perforate the bladder above the pubes without implicating its serous coat. In the female subject the uterus is almost, if not wholly, covered by the peritoneum, and the thickness of the wall between the peritoneal cavity and the vagina, at its union with the neck of the uterus, is so trifling that the former might easily be perforated, in operations upon the vagina near the os uteri.

The bladder is held in its place by five *false ligaments:* the two *posterior* being that portion of the peritoneum forming the recto-vesical fold of each side: the two *lateral,* corresponding with the obliterated hypogastric arteries and the vasa deferentia in their passage to the base of the bladder; and the *superior* being a fold of peritoneum projected between the umbilicus and the summit of the bladder by the urachus, the remains of an obliterated fœtal canal. Besides these the bladder has four true ligaments; two *anterior,* formed by the pelvic fascia reflected from the pubes to its neck and the front of its anterior surface, and two *lateral,* formed by the recto-vesical fascia, or that portion of the pelvic fascia covering the levator ani and reflected from it to the sides of the bladder.

In the female subject, the peritoneum is reflected from the uterus to the sides of the pelvis in such a manner as to form a septum between it and the bladder, which is called the *broad ligament.*

In the space between the division of the aorta into the iliac arteries, and spreading over the concave surface of the sacrum, is the *hypogastric plexus* of the sympathetic nerve, destined to the pelvic viscera; it is formed from the aortic plexus, and from branches of the lumbar nerves; it has but few ganglia, and those small ones. From this plexus originate the *hemorrhoidal, vesical, prostatic, vaginal, uterine,* and *ovarian plexuses,* supplying the parts indicated by their names, but they are only demonstrable by special dissections.

Having verified these different relations and attachments of the pelvic viscera, the next step is to trace the arteries. The bladder should be inflated, and kept so by a string tied round the penis; it is then to be drawn by hooks to the side from which the os innominatum has been removed. The divided extremity of the ureter is to be sought for, and also the vas deferens; having found and isolated these, the peritoneum is to be dissected off from the side of the pelvis, and the arteries cleared from the surrounding cellular tissue, tracing them as they are one by one given off from the internal iliac and its divisions.

The INTERNAL ILIAC ARTERY is a short trunk of large size, arising from the common iliac; it dips into the pelvis, keeping close to its walls, and divides into an anterior and a posterior trunk. The artery lies upon the internal iliac vein and the lumbo-sacral nerve; the ureter crosses it and separates it from the peritoneum. The length of the internal iliac is subject to great variation, and its branches are often irregular as to their precise point of origin, sometimes arising without the separation of the vessel into two trunks; they may all be determined by the parts to which they are distributed.

The *anterior* trunk gives off the following branches:—

>   Superior Vesical,   Obturator,
>   Inferior Vesical,   Ischiatic,
>   Internal Pudic;

and in the female subject, these two in addition:—

>   Uterine,   Vaginal.

The *hypogastric artery*, during fœtal existence, passes from the internal iliac artery, over the bladder and along the anterior parietes of the abdomen, beneath the peritoneum, to the umbilicus, and thence, with its fellow and the umbilical vein, to the placenta, forming the umbilical cord. At the close of intra-uterine life, it becomes obliterated to within an inch and a half of its commencement, leaving an impervious cord; the portion remaining pervious gives origin to the superior vesical arteries.

The *superior vesical arteries* are three or four in number, arising at intervals from the stump of the hypogastric, and are distributed to the upper part of the bladder; the most inferior of the branches is sometimes called the *middle vesical artery*.

The *inferior vesical artery* arises from the internal iliac, in common with a branch to the rectum; it is distributed to the base of the bladder, the vesiculæ seminales and prostate. The branch to the rectum is called the *middle hemorrhoidal;* it supplies the lower part of the rectum, and the vagina in the female, anastomosing with the superior and external hemorrhoidal arteries; it sometimes arises from the internal pudic.

The *obturator artery*, already referred to (p. 201), arises from the anterior division of the internal iliac; it passes forward, accompanied by the obturator nerve, which lies above it, to the upper part of the obturator foramen, beneath the horizontal branch of the pubes, where it passes out of the pelvis and divides into its terminal branches. The obturator sends a small twig to the iliacus muscle, and another to the posterior surface of the pubes.

The *ischiatic artery* is the largest branch of the anterior division of the internal iliac; it passes downward, lying upon the sacral plexus of nerves, and leaves the pelvis, just in front of the sciatic nerve, through the greater sacro-ischiatic foramen, to be distributed to the muscles of the gluteal region and the back of the thigh. Within the pelvis it gives off small branches to the rectum and base of the bladder.

The *internal pudic artery* is another branch of large size. It passes down in front of the ischiatic artery, and emerges from the pelvis at the great sacro-ischiatic foramen, crosses the spine of the ischium and enters the pelvis again by the lesser sacro-ischiatic foramen to reach the ramus of the ischium, beneath the edge of which it passes, giving off the perineal branches described at page 202, and terminating in the dorsalis penis artery.

The *uterine artery*, arising from the anterior division of the internal iliac, passes between the layers of the broad ligament and reaching the neck returns to the fundus of the uterus, giving off branches to the surface and substance of that viscus; it is usually more or less tortuous, and anastomoses freely with the ovarian artery, a branch from the abdominal aorta analogous to the spermatic artery of the male (p. 180). During pregnancy the uterine arteries increase greatly in size, and become still more tortuous. The *ovarian artery* passes between the layers of the broad ligament, and, after many tortuous convolutions, penetrates the ovary.

The *vaginal artery* seldom arises directly from the internal iliac; given off frequently from the uterine, or from the middle hemorrhoidal, it passes down in the posterior wall of the vagina to anastomose, near its termination, with the corresponding artery of the other side. Both the uterine and vaginal arteries give small branches to the bladder and rectum.

The *posterior* trunk of the internal iliac artery gives off the following branches:—

    Ilio-lumbar,   Lateral Sacral,   Gluteal.

The *ilio-lumbar artery* passes upward and then outward, beneath the external iliac, to the crest of the ilium; it there divides into two branches, one of which supplies the iliacus internus by ramifications between the muscle and the bone, forming numerous anastomoses with the circumflexa ilii from the external iliac; the other branch passes upward to supply the psoas and quadratus lumborum. This artery is analogous to the lumbar arteries; it varies considerably in its precise point of origin.

The *lateral sacral artery* passes down upon the side of the sacrum to the coccyx, inosculating with the sacra media. It sends branches through the anterior sacral foramina to the terminal portion of the spinal cord, and these, finally emerging at the posterior sacral foramina, supply the posterior surface of the sacrum.

The *gluteal artery* is a short, thick trunk, the apparent continuation of the posterior division of the internal iliac; it passes out of the pelvis at the greater sacro-ischiatic foramen, above the border of the pyriformis muscle, with the superior gluteal nerve, and is distributed to the gluteal muscles, as will be seen in the description of that region.

The *superior hemorrhoidal artery*, a branch of the inferior mesenteric, will be found distributed to the upper part of the rectum; it lies between the two layers of peritoneum constituting the meso-rectum, and anastomoses with the middle and external hemorrhoidal arteries.

All of the arteries just described are accompanied by

veins, which, with the exception of the hemorrhoidal and spermatic veins, empty into the internal iliac vein. The *hemorrhoidal* veins, more or less connected with the other veins, especially those about the neck of the bladder, empty into the inferior mesenteric vein, which terminates in the vena portæ. It will thus be seen that constipation, or any cause which obstructs the circulation of the vena portæ will also arrest the blood in the hemorrhoidal veins, and so give rise to the condition known as piles or hemorrhoids. The *spermatic* and *ovarian* veins terminate, the right in the vena cava, the left in the renal vein.

The nerves of the pelvic cavity are numerous, important, and many of them of large size.

The *lumbo-sacral* and *obturator nerves* have been already described (p. 200): they are again seen at this stage of the dissection, the former joining the sacral plexus, the latter passing out at the obturator foramen with the obturator artery.

The SACRAL PLEXUS is formed from the four anterior sacral nerves. The fifth sacral nerve terminates in the perineum, where it unites with the sixth, or coccygeal nerve, and is distributed to the side of the coccyx and coccygeus muscle. The sacral plexus is a broad, flat, nervous band lying upon the pyriformis muscle; within the pelvis, it furnishes several *visceral branches* to the pelvic organs, which unite with the branches of the hypogastric plexus, and also *muscular branches* to the internal pelvic muscles; it then divides into the following branches, destined to external parts, viz:—

    Gluteal,
    Internal Pudic,
    Great Sciatic,
    Lesser Sciatic.

These all pass out through the greater sacro-ischiatic foramen, and will be described in connection with the parts to which they are distributed.

The *pelvic fascia*, covering in the obturator internus muscle, may be traced as a single layer from the brim of the pelvis as far as a white, tendinous line stretching from the symphysis pubes to the spine of the ischium; at this line it divides into two layers, one of which continues over the rest of the obturator muscle, and the other, under the name of the *recto-vesical fascia*, passes down to be attached to the side of the bladder and rectum. Between these two layers is the levator ani muscle.

The LEVATOR ANI MUSCLE, a broad, thin plane of muscular fibres, forming, with its fellow, the floor of the pelvis, is compared by Bell to a pair of hands which dip down to hold up and support the viscera, the simile being suggested by the funnel-shaped manner in which they embrace the pelvic contents. The muscle arises from the inner surface of the os pubis, from the tendinous line constituting the point of separation between the obturator and recto-vesical fascia, and from the spine of the ischium; it is inserted into the lower part of the rectum, where its fibres become connected and continuous with those of the internal sphincter, and into the base of the bladder and the prostate. In the female, the levator ani is inserted into the side of the vagina as well as the rectum.

The two following muscles are very difficult of demonstration. They are to be sought for between the two layers of the triangular ligament, where they lie connected with the membranous urethra and pubic bones.

The COMPRESSOR URETHRÆ, or GUTHRIE'S MUSCLE, consists of two transverse layers of muscular fibres, attached by a narrow origin on each side to the ramus of the pubes; they expand at their central portion, one above and the other below the urethra, and are inserted into a fibrous raphé on the median line, extending the whole length of the membranous urethra.

WILSON'S MUSCLE is considered, when present, as a part of the preceding, and as being merely another attachment of its fibres. It arises, tendinous, from the under part of the symphysis of the pubes, and descends, fan-shaped, to be inserted into the upper layer of the compressor urethræ on the median line.

The pelvic viscera should now be removed, and in such a way as to leave the internal muscles and nerves of the pelvis uninjured; the arterial connections must be divided, and the rectum dissected up from the concavity of the sacrum. The penis should be removed with the viscera, detaching it from the arch of the pubès, by carrying the knife close to the bone. In the female, the vulva and anus should be included in an elliptical incision, and carefully dissected away from the rami of the ischia and pubes. These parts should be laid aside for further examination.

Within the pelvis will be noticed the bellies and origins

of two muscles which have their insertions outside; these are the obturator internus and the pyriformis.

The OBTURATOR INTERNUS MUSCLE arises from the bone around the obturator foramen, and from the membrane which stretches across it; it forms a triangular belly, covered by the pelvic fascia, which, tapering to a point, passes out of the lesser sacro-ischiatic foramen, to be inserted into the digital fossa of the trochanter major.

The PYRIFORMIS MUSCLE arises from the sacrum, between the first and fourth anterior sacral foramina, from the greater sacro-ischiatic ligament, and from a portion of the ilium, forming a triangular belly which terminates in a rounded tendon, and passing out at the sacro-ischiatic foramen, is inserted into the digital fossa of the trochanter major.

The COCCYGEUS MUSCLE is a small collection of muscular and tendinous fibres, arising from the spine of the ischium, and from the lesser sacro-ischiatic ligament, and inserted into the side of the coccyx; its lower border is connected with the levator ani. It is apt to be mutilated in the removal of the rectum and anus.

DISSECTION VII.

THE RECTUM.

The pelvic viscera may now be examined, commencing with the rectum; this should be cleared from all extraneous tissue, but without separating it from its connections with the bladder.

The RECTUM is about eight inches in length; it follows a curved direction, corresponding to that of the sacrum, and gradually increases in size, especially in old people, from its commencement to within an inch and a half of the anus. The last inch and a half is contracted, and follows a direction downward and backward to its termination in the anus. Its anterior surface is in contact with the bladder and its appendages in the male, and the uterus and vagina in the female, the upper portion being separated only by the recto-vesical, or recto-uterine fold of the peritoneum, while the lower portion is in direct apposition with the bladder, or separated from it only by cellular tissue. The rectum, laid open along its posterior aspect, displays a thick mucous membrane lying chiefly in longitudinal folds; at the lower part, these are called the *columns of Morgagni*, and are generally three in number. The muscular fibres are longitudinal and circular; the longitudinal cease at the lower part, and give place to the circular fibres which form the internal sphincter.

THE BLADDER.

The rectum must now be removed. The ureter and vas deferens are to be followed to their terminations. The peritoneum should be dissected from off the bladder.

The BLADDER, when distended, is of an ovoid shape, the summit or superior end being the smaller; it is connected with the penis by a somewhat funnel-shaped portion, called its *neck*. In the male, the neck is surrounded by the prostate; in the female, the place of this is supplied by cellular and muscular tissue. The summit of the bladder terminates in the *urachus*, the remains of a canal, called the allantois, which, during the early part of fœtal life, connected the bladder and the umbilical aperture.

In addition to the serous, the bladder has a muscular and a mucous coat, united together by cellular tissue. The fibres of the muscular coat are arranged both in a circular and a longitudinal manner; the circular fibres are chiefly found round the neck, and constitute what is called the *sphincter vesicæ;* the longitudinal are well marked, both in front and behind, and, from their office, are named, collectively, the *detrusor urinæ*.

The mucous coat of the bladder is thrown into folds, or becomes smooth, according to the degree of its distension. At the lower and anterior part of its interior is the orifice of the urethra, the aperture of which is partly closed by a small mucous projection, called the *uvula vesicæ*. By blowing through the ureters, their orifices will be demonstrated, as well as the obliquity with which they penetrate the bladder. The triangle formed by these two orifices, and the orifice of the bladder, is called the *trigonum vesicæ*, and is made apparent by the greater adhesion of the mucous membrane to the parts beneath than elsewhere. Especially in a hypertrophied condition of the muscular coat of the bladder, two muscular bands, proceeding from the orifices of the ureters, may be seen, on lifting the mucous coat, converging towards the urethral orifice; closely united with the sub-mucous cellular tissue, they cross each other at their point of convergence, and form the uvula vesicæ; they then become continuous with the longitudinal muscular fibres of the urethra. They are called the *muscles of the ureters*, and serve to occlude the orifice of the ureters, and to open the neck of the bladder.

MALE ORGANS OF GENERATION.[1]

VESICULÆ SEMINALES AND PROSTATE.

The vesiculæ seminales lie imbedded in a mass of cellular tissue at the base of the bladder; this is to be removed, and the prostate is also to be isolated from the veins and fascia investing it.

The VESICULÆ SEMINALES are two flattened, oblong bodies, situated upon each side of the inferior and external surface of the bladder, closely adhering to it, and converging from a point near the termination of the ureters to meet at the base of the prostate. Each vesicle consists of a coiled and sacculated tube, the convolutions of which are closely united to each other.

The VAS DEFERENS, a firm, fibrous tube, lined internally with mucous membrane, commences at the epididymis of the testicle, and accompanies the veins and arteries composing the spermatic cord, to the internal abdominal ring; it then leaves the vein and artery to pass down into the pelvis, and, getting behind the bladder, descends between it and the rectum to the inner side of the vesicula seminalis; it here becomes dilated, assumes a somewhat sacculated condition, and ends by blending with the excretory duct of the vesicula, to form the *ductus communis ejaculatorius*, which, passing through the under surface of the prostate, terminates in the urethra.

The PROSTATE is a body shaped like a chestnut, which surrounds the neck of the bladder and the commencement of the urethra; its pointed extremity is directed forward. It is invested by a dense fascia, and by a very noticeable plexus of veins called the *prostatic plexus*, which communicates with the dorsal vein of the penis, and the hemorrhoidal veins of the rectum.

The prostate is composed of *two lateral lobes*, the division of which is not always well marked. The *third lobe*, de-

---

[1] The female organs of generation are described at the close of this dissection.

scribed by Sir Everard Home, being the result of enlargement by disease of that portion of the organ situated below the urethra and behind the ducts of the vesiculæ seminales, and which the absence of resistance permits to grow out in that direction more freely than elsewhere, is, with great propriety, described by Thompson as the *posterior median portion*. The normal dimensions of the prostate are an inch and a half transversely, an inch longitudinally, and three-quarters of an inch vertically.

"The prostate consists of organic muscular fibres, arranged in a circular manner around its long axis, through which passes the urethra. It has no claim, therefore, to be regarded as a gland at all, in the sense in which that term is used to classify certain structures in the human body, but rather as a muscular body permeated by urethral glands."[1] On section, the prostate is very firm to the feel, and is of a reddish color. The orifices of its glands are numerous, and open into the prostatic part of the urethra.

THE PENIS.

The PENIS, as has been already seen, is connected to the pelvic bones by the ligamentum suspensorium and the two crura formed by the erectores penis. It is composed of the corpora cavernosa, corpus spongiosum, and glans. These are covered with integument, loosely attached by cellular tissue, that portion of it which invests the glans being called the *prepuce*.

The *corpora cavernosa* constitute the bulk of the penis; separating posteriorly, to join with the crura, they unite at the root of the penis, and are firmly connected together by a fibrous septum; they terminate anteriorly in a blunt extremity, covered in by, and closely united with, the glans. Flattened upon their superior surface, a slight groove receives the dorsal artery and vein and the dorsal nerve of the penis; these all extend to the glans. The under surface of the corpora cavernosa is more deeply grooved to receive the *corpus spongiosum;* this commences at the root of the penis, in a dilated extremity called the *bulb*, which is embraced by the acceleratores urinæ, and, continuing along the under surface of the corpora cavernosa, expands at their

---

[1] Thompson on the Prostate.

termination, to form the glans penis. The urethra occupies the centre of the corpus spongiosum.

The *glans*, being larger than the surrounding body of the penis, forms a projecting "shoulder," called the *corona glandis*, which is filled with sebaceous glands, called the *glands of Tyson*. It is covered by a delicate mucous membrane, reflected upon the inner surface of the prepuce, and continuous with the external skin of the organ. The external orifice, or *meatus* of the urethra, opens in the glans by a vertical fissure, and, from the lower end of this, the mucous membrane forms a fold between the prepuce and the gland, called the *frenum*.

The corpus spongiosum and the corpora cavernosa are invested by a dense elastic fascia, called the *sheath* of the penis. The internal structure of both these bodies and of the glans is essentially the same, being composed of what is called erectile tissue. From the inside of their investing sheaths, fibrous bands, called *trabeculæ*, passing in different directions, divide the whole interior into a multitude of minute spaces, which, upon section, have a spongy aspect, and are more or less filled with venous blood. Within these spaces is an intricate plexus of veins, completely filling them, and with which the arteries, ramifying in the trabeculæ, communicate. The blood is received from the branches of the internal pudic artery, and returned by the vena dorsalis, which, joined by short branches entering it upon the upper surface and at the root of the penis, empties into the prostatic plexus.

The *urethra* extends from the orifice of the bladder to the meatus of the glans, and is divisible into three portions: prostatic, membranous, and spongy. It should be exposed by laying it open with the scissors along the superior surface.

The *prostatic urethra* is surrounded by the prostate, the portion of which lying above the urethra is much thinner than that below. Upon its inferior wall is a small projecting crest, called the *veru montanum*, or *caput galinaginis;* the depression on each side of this is called the *prostatic sinus*, the floor of which is perforated by the orifices of the *prostatic ducts;* these may be demonstrated by squeezing the prostate; this will force out the secretion through their apertures. At the anterior part of the veru montanum is an opening, called the *sinus pocularis*, and upon the sides of this the ejaculatory, or seminal ducts, have their opening;

their orifices may likewise be demonstrated by compressing the vesiculæ seminales.

The *membranous urethra* is that portion between the prostate and the bulb; it passes through the triangular ligament, and is less than an inch in length. It is surrounded by areolar tissue and veins, and by the compressor urethræ, or Guthrie's muscle.

The *spongy portion* of the urethra, surrounded by the erectile tissue of the corpus spongiosum, extends from the bulb to the meatus. In the bulb, the urethra forms a dilatation, and at this point may be found the two orifices of the ducts of Cowper's glands; a second dilatation occurs about an inch from the meatus, and is called the *fossa navicularis*. The mucous membrane of the urethra is thin and delicate, and is sometimes thrown into longitudinal folds; it has numerous follicular orifices, one of which, in the fossa navicularis, is of large size, and is called the *lacuna magna*.

THE TESTES.

The testes are contained in an envelope called the scrotum. They may, however, remain in the abdomen, or be arrested in the inguinal canal, instead of descending from the lumbar region to the scrotum, as they should at the close of intra-uterine life. In such cases, they are usually imperfect or small, "contrary to an old authority," says John Bell; "it having been said, 'that the testicles are seated externally for chastity's sake; for such live wights as have their stones hid within their body are very lecherous, do often couple, and get many young ones.'"

The SCROTUM is composed of a tegumentary covering, and a fascia continuous with the superficial fascia of the abdomen and perineum; it is divided into two compartments by a septum, the position of which is indicated externally by a raphé continued along the under side of the penis and into the perineum. Beneath the integument is a reddish tissue, called the *dartos*, composed of non-striated muscular fibres, and in which resides the contractile power belonging to the scrotum. The testicle lies within a serous membrane, derived from the peritoneum of the abdomen, which was pushed before it in its descent during fœtal life; the connection between the portion of membrane enveloping the testicle and the peritoneum of the general cavity of the abdomen being obliterated, it forms a separate shut sac, called the *tunica vaginalis*.

In the adult, as well as in infants, there is constantly found, near the point where the testicle and epididymis become continuous, a small cellulo-fibrous and fatty body, the size of a pea, covered with serous membrane, and evidently analogous to the appendices of other serous membranes, especially of the peritoneum; it is sometimes pediculated, and at others consists merely of a little membranous fold. This is called the *appendix of the testicle*.

<small>The testicle should be examined by following out the vas deferens till it is lost in the epididymis, and then by tracing that body to its connection with the main part of the gland, removing with the scissors all the cellular tissue which surrounds it. The close adherence of the parts to each other makes the dissection a slow one.</small>

The TESTES are oval glands, suspended in the scrotum by the spermatic cord; the left is usually larger and placed lower than the right. Along its posterior side, close to that part at which the tunica vaginalis is reflected to the testicle, is a long narrow body, called the *epididymis*. The lower part of this is continuous with the vas deferens, which turns and is reflected upward beside the epididymis, being tortuous at first, but subsequently becoming straight as it unites with the other elements of the spermatic cord.

The dissection will show that besides the serous coat, the testicle has also a fibrous coat of a pearly aspect, called the *tunica albuginea*; this not only preserves the shape of the gland but sends processes into its interior for its further support; one of these, larger than the rest, lies along the posterior aspect of the gland, and is called the *mediastinum testis*.

The substance of the testicle is a pulpy mass, made up of lobules composed of the convoluted *tubuli seminiferi*, which, if seized by the forceps, may be drawn out in long threads. The lobules all converge towards the mediastinum, where their tubuli unite and, becoming larger, continue to the upper end of the testis, and finally terminate in the vas deferens. Sometimes there is an offset from the vas deferens or from the lower part of the epididymis, consisting of a prolongation of the tubuli, extending up the cord and terminating in a blind extremity; it is called the *vasculum aberrans of Haller*.

The *vas deferens* occupies the posterior part of the spermatic cord, and is easily distinguishable by its cord-like feel. Its course has been already described (p. 214).

The *spermatic cord*, besides the vas deferens, is made up of the spermatic artery and veins, and numerous lymphatics, which terminate in the lumbar glands. The spermatic veins are extremely subject to enlargement, constituting the condition known as varicocele, and always make up a very considerable portion of the cord; they may be seen commencing at the lower part of the testicle in numerous small, tortuous branches, which surround the vas deferens, and are called the *rete pampiniforme;* these, gradually enlarging, finally terminate in the single trunk of the spermatic vein, at, or near, the external abdominal ring. The cord also contains the *spermatic plexus of nerves*, which, coming from the aortic and renal plexuses, accompanies the spermatic artery. The genital branch of the genito-crural nerve and the scrotal branch of the musculo-cutaneous nerve (p. 199), are elements of the cord; and the cremaster muscle (p. 165), variably developed, the spermatic fascia and the fibro-cellular remains of the tube of peritoneum, once communicating with the abdominal cavity, also make up a large portion of its bulk.

FEMALE ORGANS OF GENERATION.

The relations of the organs peculiar to the female sex have been already studied; it remains to examine, in detail, the uterus and its appendages, including the external organs, or *vulva*. The latter will be examined first.

The prominence of the pubes, covered with a development of hair, is called the *mons Veneris*. Extending downward from this are the *labia majora*, composed of two folds of integument, meeting above and below, filled with fat and cellular tissue, and lined internally with mucous membrane; the round ligament of the uterus terminates in these. Inferiorly, upon separation, a transverse fold will be seen stretching across between them, called the *fourchette;* in women who have borne children, this is usually destroyed. Superiorly will be noticed the *clitoris*, a small projecting body, analogous to the penis of the male, composed of two corpora cavernosa, attached on each side to the ramus of the ischium, and receiving the insertions of the erector clitoridis muscle. The clitoris is composed, like the penis, of erectile tissue, and is surmounted by a glans surrounded by a prepuce. From this prepuce, a longitudinal fold of mucous membrane descends on each side, and becomes blended with the labia majora; these folds are called the *nymphæ*, or *labia minora*. Between these, and just above the aperture of the vagina, is the orifice of the urethra, surrounded by an elevated margin. The urethra is about an inch and a half in length, and lies in the upper wall of the vagina, from which it cannot be separated; it is very elastic, and capable of great distension. The orifice of the vagina is transversely elliptical; in the virgin it is sometimes partly closed by a circular fold of mucous membrane, called the *hymen;* this is destroyed by sexual intercourse, or by child-birth, but its former presence is indicated by small elevated excrescences, called *carunculæ myrtiformes*. Just anterior to the hymen may be found the orifices of two ducts, one on each side, often made apparent from being distended with sebaceous matter; by laying these open, each may be traced to a round body, the size of a large pea, called the *gland of Bartholinus*.

The parts just described are supplied by branches of the internal pudic artery, and by offsets from the lumbar and sacral plexuses of nerves.

The bladder may now be dissected from the uterus, and the vagina laid open with the scissors along its superior surface.

The *vagina* occupies a position corresponding to the axis of the outlet of the pelvis, and reaches from the cervix of the uterus, which projects into it, to the external opening of the vulva; it is placed between the bladder and the rectum, and at its sides is embraced by the levatores ani muscles. Its orifice is surrounded by a sphincter muscle (p. 203), and its upper extremity is dilated; its internal surface is lined with mucous membrane, thrown into rugæ, more marked near the entrance than higher up; these meet in a raphé which extends along the centre of both the anterior and posterior walls; the two raphés being called the *columns of the vagina*. Beneath the mucous membrane is a layer of contractile tissue similar to the dartos, and external to this a layer of cellular tissue; the upper part of the posterior wall of the vagina is covered by the peritoneum of the recto-uterine fold.

The UTERUS, with the exception of its mouth and neck, is covered with peritoneum, which spreads out on both sides into what is called the *broad ligament*, stretching across the pelvis, as already seen, and forming a sort of septum between the bladder and rectum. It is pyriform in shape, convex posteriorly and flattened anteriorly, and is divided into a fundus, body, cervix, and os.

The uterus may be laid open with the scissors, introduced at the os, making a longitudinal incision to be afterward crossed by a transverse one at the fundus.

The substance of the uterus is composed of muscular fibres. In its natural state, the muscular portion is seen as a firm, compact, fibrous tissue; in the impregnated condition, this becomes hypertrophied and vastly more apparent. The orifices of divided veins will be seen on the face of the section; these, in the pregnant uterus, are called the *sinuses*.

The *cavity of the uterus* is lined with mucous membrane continuous with that of the vagina; it is triangular in shape, and about the size of an almond; the base of the triangle corresponds to the fundus of the organ, the *fundus* being the broad portion surmounting the *body*, or central part. The cavity is constricted at the union of the body with the cervix, the *cervix* being the portion between this constriction and the mouth, or os uteri. The point of constriction between the cervix and body is sometimes called the *os uteri internum*. The canal of the cervix is slightly dilated, and the oblique folds of the mucous membrane, in this part, have received the name of *arbor vitæ uteri*. The *os uteri*, before parturition, consists simply of a rounded orifice with thick lips; after childbirth it becomes a transverse fissure, the posterior lip of which is the longest.

The uterus is supplied with blood from the uterine and ovarian arteries. The uterine veins are large and numerous, and form plexuses upon each side of the organ; its nerves are derived from the hypogastric and ovarian plexuses, and from branches of the third and fourth sacral nerves.

The *appendages of the uterus* are the broad ligaments, the round ligaments, the Fallopian tubes, and the ovaries.

The *broad ligaments*, consisting simply of the two peritoneal layers covering the anterior and posterior uterine surfaces, and then reflected upon the walls of the pelvis, have been already referred to.

The *round ligaments* are round cords of fibrous tissue attached to the sides of the fundus uteri, which, passing upward and outward, to the internal inguinal ring and through the inguinal canal, like the spermatic cord in the male, are lost in the mons Veneris and labia majora.

The *Fallopian tubes*, about four inches in length, are each connected by one end with the side of the fundus of the uterus; the other end is loose in the cavity of the pelvis. Their canals are very minute, but traceable with a fine probe to the openings at the angles of the cavity of the uterus, with the mucous membrane of which their own becomes continuous. The outer termination of the tube is free in the peritoneal cavity; it is dilated and surrounded by a circular, fringe-like fold, called the *corpus fimbriatum*, or *morsus diaboli*. It will be noticed that there is thus a communication between the cavity of the uterus and that of the peritoneum, forming a single exception to the general rule that serous membranes are shut sacs.

The OVARIES are two white, oblong bodies, with either a smooth or scarred surface, bulging from the posterior aspect of the broad ligaments, and with which they are connected by their anterior margins. They are attached to the uterus by their inner extremity, through the medium of rounded cords called the *ligaments of the ovaries*. The ovary consists of a fibrous structure containing small *vesicles*, named *Graafian*, and the peritoneum surrounds the whole organ, except at its attached border. Their dimensions are very variable, but may be stated, in a general way, as an inch and a half, by three-quarters of an inch in diameter, and half an inch in thickness.

The ovaries are supplied by the ovarian arteries, which anastomose with the uterine. The ovarian plexus of nerves, derived from the aortic plexus, accompanies the ovarian artery, and the uterine nerves are also, in part, distributed to them.

It is not easy for the student to tell which is the front and which the back of the uterus; the ovaries, when present, will tell him; they being always placed on the posterior aspect of the broad ligament; in their absence, he must remember that the uterus is convex posteriorly, and flat anteriorly.

DISSECTION VIII.

ANTERIOR FEMORAL REGION.

In studying the anatomy of femoral hernia, the superficial structures of the anterior femoral region have, in part, been examined. The incision of the skin may now be carried down the front of the thigh to the knee, and the integument reflected; several cutaneous nerves will be exposed by this process.

The CUTANEOUS NERVES are branches of the external cutaneous, genito-crural and crural nerves, all being branches of the lumbar plexus. The *external cutaneous* pierces the deep fascia just below the anterior superior spinous process of the ilium, and is distributed to the integument of the anterior part of the gluteal region and the outside of the thigh. The *crural* branch of the genito-crural nerve is small in size, and, emerging from the sheath of the femoral artery, is distributed to the anterior aspect of the thigh. The *crural nerve*, appearing just below Poupart's ligament from between the psoas and iliacus muscles, divides into two branches. One of them, deeper than the other, subdivides to supply the muscles on the fore part of the thigh and the pectineus muscle on the inside. The other division is composed of cutaneous branches, distributed to the integument of the anterior aspect of the thigh as far down as the patella. The largest of these branches is called the *long saphena nerve;* this accompanies the femoral artery on its outer side, and at the opening in the tendon of the adductor magnus muscle, leaving the artery and passing beneath the sartorius muscle, becomes subcutaneous and descends the inner side of the leg to the inner border of the foot.

The *internal saphena vein*, especially if distended with blood, will be easily traced, in its superficial course, from the inside of the foot, along the inner border of the leg, behind the inner condyle, upward to the saphenous opening. This vessel affords an excellent opportunity to examine the *venous valves;* they are formed from the internal lining membrane, and are usually in pairs, and of a semilunar shape; their situation is indicated externally by the dilatation of the vessel above them.

Upon the outside of the thigh the deep fascia, or fascia lata, will be found to become more aponeurotic, and, at its upper part, to divide into two layers, between which lies the tensor vaginæ femoris.

The TENSOR VAGINÆ FEMORIS is a short and thick muscle, arising from the anterior superior spinous process and a portion of the crest of the ilium; it is inserted into the fascia lata between its two layers, at a point five or six inches below that of its origin.

<small>The fascia lata may be removed as far as the anterior border of the tensor vaginæ muscle, but it should not be divided transversely until the vastus externus is dissected. The removal of that part of the fascia in front of the thigh exposes the sartorius and rectus muscles. In dissecting the sartorius it will be found advantageous to remove its anterior sheath without disturbing the posterior adhesions which keep the muscle in place: when this is accomplished it may be dissected posteriorly, but it is so long a muscle that it is difficult to keep it tense, or to dissect it neatly.</small>

The SARTORIUS MUSCLE arises from the anterior superior spinous process of the ilium, and crosses the thigh obliquely to the inside of the knee, where it forms a thin, flat tendon which is inserted into the inner tuberosity of the tibia by an aponeurotic expansion, which loses itself in the fibrous tissues surrounding the knee-joint.

<small>In dissecting the origins of the three following muscles, they must be free from all tension; but in preparing their bellies and insertions they should be made tense, and this will be accomplished by bending the knee upon the thigh. The length and flaccidity of the sartorius permit its being drawn to one side, so as not to interfere with the dissection of the parts beneath. The second head of the rectus is not easily seen, as it is very short; by lifting the muscle upward and inward it may be isolated from the surrounding parts, and brought into view.</small>

The RECTUS FEMORIS MUSCLE lies upon the front of the femur, forming a beautiful muscular mass, the fibres of which radiate from a central longitudinal line in a bipenniform manner. It arises by two tendinous heads, one from the anterior inferior spinous process of the ilium, just below the origin of the sartorius; the other from the upper surface of the acetabulum; it is inserted by a broad flat tendon into the upper border of the patella.

The VASTUS EXTERNUS MUSCLE is partly covered in by the fascia lata into which the tensor vaginæ femoris is inserted; it forms the bulk of the outer, fleshy part of the thigh; it arises from the base of the trochanter major, and from the whole length of the linea aspera, and is inserted into the outer side of the patella, forming a common aponeurotic tendon with the rectus muscle.

The VASTUS INTERNUS MUSCLE forms the bulk of the fleshy portion of the lower part of the thigh at its inner side, giving the limb its characteristic outline; it arises from the anterior and lateral surfaces of the femur, and from the whole length of the linea aspera; its lower end terminates in an aponeurosis, which, blending with that of the rectus, is inserted into the inner border of the patella. The upper part of the muscle is hidden beneath the rectus and sartorius muscles; the adductor muscles are inseparably connected with that portion attached to the linea aspera. The portion lying beneath the rectus muscle, the fibres of which run longitudinally from the inter-trochanteric line to the patella, is sometimes described as a separate muscle, called the *cruræus*. The *sub-cruræus muscle* is a small bundle of fibres beneath the cruræus, arising from the front part of the femur, and inserted into the capsule of the knee-joint.

These three muscles are sometimes described as one muscle with three heads; the patella is then considered as a sesamoid bone, and the ligamentum patellæ, inserted into the tubercle of the tibia, as the real tendon of insertion to the three muscles, combined under the name of *triceps extensor cruris*. If the cruræus is looked upon as a separate muscle, the term *quadriceps* is used.

These muscles are all supplied by branches of the crural nerve, and the femoral artery sends off irregular muscular twigs, which penetrate them on their under surface.

The muscles upon the inner and upper part of the thigh almost invariably become dried and defaced from the delay occasioned by the examination of other parts, after their exposure in the dissection of femoral hernia; if they have been kept properly moist, they will however still be found in tolerable condition.

If the subject be entire, the knees should be bent and the soles of the two feet placed in contact with the heels pushed up and so approximated to the pelvis that they will retain their position. If the limb is separated from the trunk, the ilium must be fixed by blocks, and the muscles of the inside of the thigh rendered as tense as circumstances will permit.

The muscles of the inner side of the thigh are the three adductors (longus, brevis, and magnus), with the gracilis and pectineus. The gracilis is the longest and most internal; superficial to the others are the pectineus and the adductor longus, and beneath these the adductors brevis and magnus.

The GRACILIS MUSCLE arises by a thin, flat, and broad tendon from the side of the symphysis pubes and from the ramus of the pubes and ischium; it forms a long, ribbon-like belly, which, passing down on the inside of the thigh, terminates in a tendon, rounded at first, but becoming flattened, and is inserted into the head of the tibia, beneath the expanded insertion of the sartorius.

The ADDUCTOR LONGUS MUSCLE consists of a large fleshy belly, arising by a round, tendinous origin from the front of the os pubis, and inserted into the middle third of the linea aspera by an aponeurosis which is partially confounded with that of the adductor magnus. The dissection of this muscle at its upper part will expose the profunda artery, and at its insertion the femoral artery will be seen in close relation to it.

Between this muscle and the femur lies the PECTINEUS, arising by a broad and flat muscular origin from the ilio-pectineal line, and inserted into the ridge leading from the lesser trochanter to the linea aspera. It is not easy to find the line of separation between this muscle and the conjoined tendon of the psoas and iliacus, the insertion of which into the lesser trochanter may now be seen (p. 200).

The TRIANGLE OF SCARPA is the triangular depression in the upper part of the thigh, the base of which is formed by Poupart's ligament, the outer side by the inner border of the sartorius, and the inner side by the superior border of the adductor longus. The femoral artery runs through the centre of this space, with the femoral vein internal to it; half an inch external to the artery is the crural nerve, at first deep-seated between the psoas and iliacus, afterward becoming more superficial.

The FEMORAL ARTERY extends from Poupart's ligament to the point at which it perforates the adductor magnus muscle; it is covered in by a strong sheath, common to it and the vein, and occupies the depression existing between the adductor muscles on the inside of the thigh and those which cover the femur upon the outside; it lies upon the psoas, pectineus, and adductor longus muscles, and the femoral vein, except at the upper part, where it is upon its inner side, lies almost directly behind it; it is accompanied by the long saphena nerve. The artery disappears through an aperture in the tendon of the adductor magnus muscle, at the union of the middle and lower third of the thigh, and which is the commencement of a fibrous canal, formed by the tendons of the adductor magnus and vastus internus muscles, called *Hunter's canal;* the sartorius muscle, usually called the satellite of the femoral artery, crosses the thigh at such constantly varying angles, that its precise relation to it is by no means constant. The artery may sometimes be found split into two trunks below the origin of the profunda; these, however, always unite before the vessel perforates the adductor magnus. The point of election for applying a ligature to the femoral artery, is, the thigh being rotated outward, at the apex of Scarpa's triangle, where the sartorius muscle crosses the vessel; it corresponds to the bisection of a line drawn from the centre of Poupart's ligament to the posterior edge of the inner condyle of the femur, by a transverse line indicating the upper fourth of the thigh.

The superficial branches given off by the femoral artery have been already dissected and described, at p. 195, in connection with the anatomy of femoral hernia.

At a variable distance, viz: from one half to two inches below Poupart's ligament, the femoral artery gives off, posteriorly, a large branch, nearly equal in size to the main trunk itself, and called the *profunda artery.* This, passing backward behind the adductor longus, breaks into branches which perforate the adductor muscles, and are distributed to the parts on the back of the thigh.

The dissection of the profunda is with difficulty accomplished, without destroying parts yet to be dissected, owing to the number of its branches, the confined limits of the space they occupy, and the depth to which they penetrate. The scissors will be found useful at this time, and, by the aid of hooks and by a judicious position of the limb, the soft fat and cellular tissue surrounding the arteries may be removed, and the branches, if not too much meddled with by the forceps, neatly displayed.

The profunda gives off an external and an internal circumflex artery; one or both of these sometimes arise directly from the femoral.

The *external circumflex* is the larger of the two; it passes outward beneath the rectus muscle and divides into three branches, or sets of branches, viz: *ascending,* to inosculate with the gluteal artery on the dorsum of the ilium, near its crest; *middle,* which curve around the femur, just beneath the greater trochanter, to inosculate with the gluteal and internal circumflex arteries; and *descending,* which are distributed to the muscles of the outside of the thigh.

The *internal circumflex* passes beneath the heads of the adductor muscles on the inside of the thigh, and is only to be farther traced by their division; it supplies these muscles, sends a branch to the hip-joint, and inosculates with the external circumflex, obturator, and ischiatic arteries.

The terminal branches of the profunda are called the *perforating arteries*, because they pass through foramina in the adductor tendons to reach the back part of the thigh. They are usually three in number; they anastomose freely with each other, with the ischiatic and internal circumflex arteries above, and with the articular branches of the popliteal artery inferiorly. One of these arteries supplies the *nutrient branch* to the femur.

These branches of the femoral artery are all accompanied by veins which unite to form the *profunda vein;* this enters the femoral vein an inch or more below Poupart's ligament.

The continuation of the femoral artery furnishes muscular branches to the muscles contiguous to it, and as it is about to perforate the adductor magnus tendon, gives off a branch called the anastomotica magna.

The *anastomotica magna* is not usually of large size, nor always constant in its point of origin, not unfrequently arising from the popliteal artery; it is accompanied by the long saphena nerve, and descending to the inner condyle, inosculates with the superior internal articular branch of the popliteal. In very finely injected subjects, numerous anastomoses may be traced between this artery and other branches distributed to the neighborhood of the knee-joint.

The pectineus and adductor longus muscles may now be divided in the middle, and their ends reflected. The arteries should be preserved so far as possible.

In removing these muscles, the *obturator nerve* and its branches should be sought for; emerging from the pelvis (p. 199) it is distributed to the muscles of the inner side of the thigh, and a long branch passing behind the pectineus, descends to the knee, where it joins with branches of the long saphena nerve.

Small branches of the obturator artery will also be found at the upper part of these muscles.

The ADDUCTOR BREVIS MUSCLE lies immediately beneath the two muscles just divided; arising by a narrow origin from the external surface and ramus of the os pubis, and passing very obliquely inward, it is inserted by a broad tendon, behind the pectineus, into the upper part of the linea aspera.

The ADDUCTOR MAGNUS MUSCLE, the deepest of the three

adductors, forms the bulk of the upper part of the thigh, and separates the muscles of the anterior and posterior femoral regions; superiorly it lies beneath the adductor brevis, which should consequently be removed. It arises from the rami of the pubes and ischium, and from the tuberosity of the ischium, and is inserted into the intertrochanteric line, and the whole length of the linea aspera, as far as the inner condyle, where it terminates in a rounded tendon. At its lower part it becomes confounded with the adductor longus and the vastus internus muscles. At its upper part it is pierced by the perforating arteries, and lower down it has a large, oval, and tendinous opening, converted into a canal by the tendon of the vastus internus muscle, called *Hunter's canal*, and through which pass the femoral vessels.

By detaching the origin of the adductor magnus, the obturator externus muscle will be exposed.

The OBTURATOR EXTERNUS MUSCLE arises from the rami of the pubes and ischium, and from a part of the surface of the obturator membrane; it forms a triangular belly, and its fibres converge to a rounded tendon, which is inserted into the digital fossa of the great trochanter.

The *obturator artery* (p. 208), after emerging from the pelvis, divides into two branches, one of which forms a circle around the membrane beneath the obturator muscle, and sends an *articular* twig, through the notch of the acetabulum, to the head of the femur, which it reaches by means of the round ligament; the other branch supplies the obturator and adductor muscles, and unites with the internal circumflex artery.

DISSECTION IX.

GLUTEAL REGION.

The subject should be turned over, and a high block placed beneath the thighs, in such a way that the pelvis may hang over its edge, and yet remain fixed firmly enough for dissection; the thigh should be rotated inward, and the foot should lie upon its outer side. An incision is to be made obliquely outward from the upper part of the sacrum, to a point a hand's breadth below the greater trochanter; this should penetrate to the muscle, and the flaps be reflected by dissecting in the direction of its fibres. The skin is thick and tough, and there is a deep layer of fat beneath it, which will probably be found more or less infiltrated with the fluids which have gradually gravitated into the part while dependent.

The gluteus maximus, which is the first muscle exposed, is composed of coarse bundles of fibres, between which penetrate prolongations from its sheath. It is one of the most difficult muscles in the body to dissect neatly, though in a favorable subject, and when well dissected, few present a more showy appearance.

The GLUTEUS MAXIMUS MUSCLE arises from the posterior fifth of the crest of the ilium, and of the bone beneath, from the lateral tubercles of the sacrum, the sacro-iliac and the greater sacro-ischiatic ligaments, and from the side of the coccyx; its upper half is inserted into a thick, flat tendon, covering in the greater trochanter and continuous with the fascia lata of the thigh; the lower portion is inserted into the rough line on the femur, leading from the trochanter major to the linea aspera. Beneath this broad tendon is a large synovial bursa. The upper border of this muscle is closely adherent to the aponeurotic exterior of the gluteus medius muscle, by means of the fascia lata; its lower border forms the fold of the nates, and overlaps the origins of the muscles of the back of the thigh.

This muscle is to be divided transversely, and its two ends are to be reflected; this will expose the gluteus medius.

In dividing the gluteus maximus, a number of arterial twigs, distributed to the muscle, will be cut across; they are muscular branches of the gluteal and ischiatic arteries; a few ascending twigs from the external circumflex will also be seen.

The LESSER SCIATIC NERVE, coming from the sacral plexus, emerges with the ischiatic artery at the sacro-ischiatic foramen. Its branches supply the integument and the gluteus maximus, which it penetrates at its lower border; some of its branches descend and are distributed upon the posterior aspect of the thigh; one of these, larger than the others, is called the *middle posterior cutaneous nerve*.

The GLUTEUS MEDIUS MUSCLE, partially covered by the gluteus maximus, and elsewhere by a dense aponeurosis, which cannot be dissected from the muscular tissue beneath, arises from the anterior four fifths of the crest of the ilium, and from the superior curved line of the external surface of that bone, being closely connected at its anterior border with the gluteus minimus and the tensor vaginæ femoris; its fibres converge to be inserted by a thick tendon into the external surface of the trochanter major. The gluteal artery emerges at its posterior border, and a large branch ramifies upon this muscle, between it and the gluteus maximus.

This muscle may now be removed, by dividing it an inch above its insertion, and reflecting its muscular belly. A confused mass of areolar tissue, arteries, nerves, and muscles, remains, which is to be patiently cleared up, by gradually removing the areolar tissue with the scissors and forceps, following out the arteries and nerves; the tracing of them should be commenced at the point at which they emerge from the pelvis.

The *gluteal artery* emerges between the gluteus medius and pyriformis muscles, and is, as has been stated (p. 209), a branch of the internal iliac, the terminal one of its posterior division; it breaks up into branches as soon as it emerges from the pelvis. The *superficial branch* goes to the gluteus maximus; the *deep superior* passes upward and forward to the anterior superior spinous process of the ilium, between the gluteus medius and minimus muscles, and inosculates with the circumflexa ilii and external circumflex arteries; the *deep inferior* branch ramifies upon the gluteus minimus, in the neighborhood of the trochanter and capsule of the hip-joint.

The *gluteal nerve*, coming from the sacral plexus, emerges with the gluteal artery, and divides into two branches; one, passing upward toward the crest of the ilium, and supplying the gluteus medius and minimus; the other, passing forward, sends branches to these muscles, and terminates in the tensor vaginæ femoris.

The *ischiatic artery*, one of the terminal branches of the anterior division of the internal iliac, emerges below the pyriformis muscle, and passes downward between the trochanter and the tuberosity of the ischium, supplying muscular branches to the gluteus maximus and posterior muscles of the thigh; it also sends a branch to the great sciatic nerve, named *comes nervi ischiatici*, which accompanies it to the popliteal space.

The GREAT SCIATIC NERVE is the continuation of the sacral plexus. It emerges below the pyriformis muscle as a broad, flat cord, three-fourths of an inch in width; it descends between the trochanter and tuberosity of the ischium, and will be further seen in the dissection of the posterior region of the thigh. It supplies the heads of the posterior muscles of the thigh, and gives off some small branches which ramify on the capsule of the hip-joint. Not unfrequently it is split into two trunks, one of which passes through the pyriformis; sometimes the whole nerve perforates that muscle.

The *internal pudic artery* emerges below the pyriformis muscle, in front of the ischiatic artery, at the great sacro-ischiatic foramen, passes under the greater sacro-ischiatic ligament, and ascends along the ramus of the ischium; it is accompanied by the internal pudic nerve; its branches and distribution have already been described (p. 205).

In dissecting these arteries, the following muscles will have been exposed.

The GLUTEUS MINIMUS MUSCLE arises, fan-shaped, from the surface of the ilium, between the acetabulum and the middle curved line of that bone, and is inserted into the summit and inside of the great trochanter; anteriorly this muscle is blended with the gluteus medius.

The PYRIFORMIS MUSCLE, arising from the sacrum, has already been described as seen within the pelvis (p. 212); the part external to the pelvis will now be found narrowing to its insertion and separated from the gluteus minimus by the gluteal artery. It arises from the concave surface of the sacrum, between the first and fourth anterior sacral foramina, from the greater sacro-ischiatic ligament, and from a portion of the ilium; forming a thick, flattened belly, it passes out at the great sacro-ischiatic notch, tapers to a rounded tendon, blending with that of the gluteus minimus, and is inserted into the digital fossa of the great trochanter.

Some little difficulty is often experienced by the student in determining the three next muscles. A bundle of muscular fibres presents itself just below the pyriformis, and crossed by the great sciatic nerve; it is composed of the tendon of the internal obturator muscle surrounded by its two dependent muscles, the gemelli; if this bundle of fibres be carefully separated longitudinally, the glistening tendon of the obturator will be exposed, and the two gemelli may be distinctly defined, on one side and the other, though they cannot be isolated from the obturator tendon.

The OBTURATOR INTERNUS MUSCLE has been partly seen in another dissection (p. 212); it arises within the pelvis from the margin of the bone which surrounds the obturator foramen, and from the membrane which stretches across it; it passes over the lesser sacro-ischiatic notch, which acts as a pulley on which its tendon plays in the change of direction which it assumes to reach its insertion. The belly of the muscle is broad and flat, and, tapering to a rounded tendon, embraced by the gemelli muscles, is inserted into the digital fossa of the great trochanter. On dividing this muscle, it will be seen that the ischiatic notch is covered with cartilage and provided with a synovial bursa.

The GEMELLUS SUPERIOR arises from the spine of the ischium; the GEMELLUS INFERIOR from the upper and posterior part of the tuberosity of the ischium. The superior is usually the smaller of the two, and they embrace, and either wholly or partially conceal, the tendon of the obturator muscle, into which some of their fibres are inserted, while the rest are inserted with that tendon into the digital fossa of the greater trochanter. These two muscles thus constitute appendages of the obturator internus, and form, as it were, a "marsupium," or pouch, to that muscle, which, from having connected with it a well-marked bursa, was sometimes called, by the old anatomists, the "*bursalis*," or "*marsupialis*" muscle. The gemellus superior is sometimes absent.

The QUADRATUS FEMORIS MUSCLE lies next below the gemellus inferior; it is a flat and quadrangular muscle, arising from the external border of the tuberosity of the ischium, and inserted into the linea quadrati at the posterior and lower part of the greater trochanter. This muscle is tendinous at its insertion, and its lower border is in relation with the adductor magnus; it is crossed by the great sciatic nerve, and the internal circumflex artery emerges at its upper border. Just above this muscle will be seen the tendon and part of the belly of the obturator externus muscle, going to be inserted into the digital fossa.

DISSECTION X.

POSTERIOR FEMORAL REGION.

An incision is to be carried down the back of the thigh, a short distance below the fold of the knee-joint, and the integument reflected.

Upon the fascia lata, thus exposed, will be found a number of cutaneous nerves, on the inside, derived from the obturator nerve, on the outside from the external cutaneous nerve (p. 199), and in the middle from the middle posterior cutaneous branch of the lesser sciatic nerve (p. 230). At the lower part of the thigh, the external saphena vein, ascending from the foot along the median line of the calf of the leg, may be seen penetrating the popliteal space to join the popliteal vein.

The fascia lata is to be removed and the muscles are to be dissected. The sciatic nerve and its divisions are to be particularly respected.

The posterior femoral muscles will be found supplied with arteries from the circumflex and perforating branches of the profunda artery.

The BICEPS MUSCLE is the most external of this region, and, in common with the two muscles to be next described, is covered in at its upper part by the gluteus maximus. It arises from the tuberosity of the ischium by a tendon, only artificially separable from the other muscles arising at that point, and also from the femur by a second head attached to the linea aspera; these two heads unite to be inserted by a round tendon into the head of the fibula and outer tuberosity of the tibia. This tendon forms the outer hamstring.[1]

The SEMI-TENDINOSUS MUSCLE lies upon the inner side of the posterior femoral region. At its origin from the tuberosity of the ischium, and for some distance below it, it is not easily separable from the biceps; it forms a comparatively short and stout belly, terminating in a long tendon, which is inserted into the inner surface of the tibia below the tendon of the gracilis, both of these tendons being covered in by the expanded insertion of the sartorius.

The SEMI-MEMBRANOSUS MUSCLE lies beneath the two preceding muscles, and derives its name from the membraniform tendon which characterizes its origin. It arises from the tuberosity of the ischium, in common with the biceps and semi-tendinosus muscles, and is inserted by a tendon which has three different points of attachment, viz: an internal, to the inner tuberosity of the tibia; a middle

---

[1] The student can assist his memory to retain the fact that the biceps forms the outer hamstring, by the first two letters in the word *Boston*. (B. O. biceps, outer.)

which is continuous with the fascia covering the popliteus muscle, and a posterior, which expands upon the posterior surface of the knee-joint and is attached to the outer condyle of the femur, constituting what is called the ligamentum posticum Winslowii of the knee-joint. The tendons of the semi-membranosus and semi-tendinosus form the inner ham-string, and with that of the gracilis, from a fancied resemblance, derived from their divergence, have received the name of the *pes anserinus*.

The SCIATIC NERVE, surrounded by a considerable amount of fat and areolar tissue, will have been traced in the foregoing dissection (p. 231). At the upper part of the thigh it is comparatively superficial, there being no muscle between it and the integument; it then passes underneath the long head of the biceps, and down upon the outer side of the median line of the thigh to the popliteal space, giving off, in its course, branches to the muscles of the posterior femoral region, between which it lies, and a single articular branch to the knee-joint. Toward the lower part of the thigh, it divides into the popliteal and peroneal branches; sometimes this division takes place higher up, even before emerging from the pelvis; in which case, as has been seen, one of the branches perforates the pyriformis muscle.

POPLITEAL SPACE.

The POPLITEAL SPACE is the diamond-shaped interval between the biceps and semi-tendinosus and semi-membranosus muscles above, and the separated heads of the gastrocnemius muscle below; its base, or floor, being the flat surface of the femur above the condyles; the fascia lata and integument cover it superficially, and it is traversed by the popliteal artery and vein, and the popliteal nerve with its branches. The relative position of these is as follows: The popliteal nerve is the most superficial, and its situation corresponds to the long diameter of the popliteal space; immediately beneath the nerve is the popliteal vein, and directly under the vein, the popliteal artery, the nerve, vein, and artery lying superimposed one upon another. Between the nerve and the vein there is usually an interval filled with fat.

The branches of the sciatic nerve which are superficial are to be examined first.

The *peroneal* division of the sciatic nerve is smaller in size than the popliteal; it accompanies the tendon of the biceps muscle to the head of the fibula, where it curves around that bone, passing between it and the peroneus longus muscle, dividing, as it disappears, into the anterior tibial and musculo-cutaneous nerves. The peroneal nerve, before disappearing, besides giving off a few small cutaneous, muscular, and articular branches, furnishes an important superficial branch, the *communicans peronei;* this passes down upon the outer side of the calf of the leg to about its middle, where it is joined by a branch from the popliteal, called the communicans poplitei, and the two, uniting, constitute the external, or short saphena nerve. This union does not always, however, take place, one of the nerves losing itself in the integument, and the other, continuing downward, pursues the course and takes the name of the short saphena nerve.

The *popliteal nerve* lies superficial to the popliteal vein, and, at the lower border of the popliteal space, becomes the posterior tibial nerve; while in the space, it gives off *muscular* branches to the heads of the gastrocnemius muscle, an *articular* branch to the interior of the knee-joint, and a superficial branch, called the *communicans poplitei;* this is larger than the communicans peronei, with which, after passing between the two heads of the gastrocnemius, it unites, half-way down the leg; their union constitutes the *external*, or *short saphena nerve*. This nerve descends the leg on the outer side of the tendo Achillis, curves round the outer malleolus, and is distributed to the outside of the foot and little toe.

The POPLITEAL VEIN lies upon the popliteal artery, and receives the external saphena vein, which, after a superficial course up the back of the leg, here penetrates to terminate in the popliteal vein.

The POPLITEAL ARTERY lies deeply imbedded in the popliteal space; it is the continuation of the femoral artery; commencing at the point where that vessel emerges, after passing through the opening in the tendon of the adductor magnus muscle, it terminates at the lower border of the popliteus muscle, by dividing into the anterior and posterior tibial arteries; this division sometimes takes place at a point higher up.

The popliteal artery gives off four articular branches, viz: *superior* and *inferior external articular*, and *superior* and *inferior internal articular*. These wind around the knee-joint to its front, supplying the lower part of the femur, the heads of the tibia and fibula, and the joint itself; they anastomose very freely with each other, with the anastomotica magna, and the recurrent branch of the anterior tibial artery. The inferior articular arteries pass beneath the lateral ligaments of the knee-joint, and the superior articular beneath the tendon of the biceps on the outside, and that of the adductor magnus muscle on the inside. Two large muscular branches, called the *sural*, pass downward, to supply the heads of the gastrocnemius muscle, and the *azygos articular*, a small branch, almost always broken or destroyed in removing the fat from the popliteal space, springing from the posterior aspect of the artery, penetrates the ligamentum posticum Winslowii, to supply the internal structures of the knee-joint.

DISSECTION XI.

FRONT OF THE LEG AND DORSUM OF THE FOOT.

The dissection now reverts to the front of the limb, and is continued by making an incision from the knee downward and along the dorsum of the foot. The integument alone should be reflected.

The front of the leg will be found encased by a strong, glistening fascia, continuous with the fascia lata of the thigh; upon the inner side of this will be found the internal saphena vein, and, at the junction of the middle and lower third, the continuation of the musculo-cutaneous branch of the peroneal nerve will be seen emerging through an aperture in the fascia.

The *internal saphena vein* commences at the great toe, and passes up along the inner side of the leg, and behind the inner condyle, to the saphenous opening in the upper part of the anterior femoral region (p. 196), where it terminates in the femoral vein.

The *musculo-cutaneous nerve*, after its division from the peroneal at the head of the fibula, passes downward between the peronei muscles and the extensor longus digitorum, to emerge through a distinct foramen in the fascia on the front of the leg, at about the union of its middle and lower third; it then divides into two branches, called the *peroneal cutaneous;* these pass down superficially, and at the toes divide again into branches, distributed to their sides, except the outer side of the little toe.

On the dorsum of the foot, the termination of the *anterior tibial nerve* (p. 240) will be found, accompanying the dorsalis pedis artery in the first interosseous space; this nerve generally supplies the great toe, and the outer side of the next;·it anastomoses with the peroneal cutaneous nerves.

Along the outer border of the foot will be found the

termination of the *external saphena nerve*, which supplies the little toe, and sometimes the outer side of the next. The distribution of the nerves to the toes constantly varies.

The *fascia of the leg*, extending across from the anterior surface of the tibia to the fibula, is extremely dense, and, at the upper part of the leg, binds down and is closely adherent to the muscles beneath, some of their fibres originating from it. At the ankle, it forms the *anterior annular ligament;* this is a thickened portion, crossing from the external malleolus and upper surface of the os calcis to the internal malleolus, and the borders of which are imperfectly defined; it contains three sheaths for the tendons of the muscles which pass down from the leg to the dorsum of the foot, viz., one for the tibialis anticus, one for the extensor proprius pollicis, and the third for the extensor longus digitorum and peroneus tertius muscle. The under portion of the sheath for the extensor proprius pollicis muscle converts that part of the ligament into a sort of sling, and is attached to the os calcis under the name of *ligament of Retzius.*

The fascia of the leg should be removed by detaching it from below upward; where it becomes blended with the bellies of the muscles, it should be left in connection with them. The annular ligament is not to be removed. The muscles of this region should be separated from each other by following them up from their tendons.

The TIBIALIS ANTICUS MUSCLE lies along the side of the tibia, arising from the upper two thirds of its inner surface and from the interosseous membrane; its tendon passes through a separate sheath in the annular ligament, and is inserted into the side of the internal cuneiform bone and the head of the first metatarsal bone.

The EXTENSOR PROPRIUS POLLICIS is a thin muscle, covered in by the tibialis anticus on one side and the extensor longus digitorum on the other; it arises from the middle third of the fibula, and its tendon, which runs along the anterior border of the muscle, receiving the fibres in a penniform manner, passes through a separate sheath in the annular ligament, and is inserted into the base of the second phalanx of the great toe.

The EXTENSOR LONGUS DIGITORUM arises from the upper half of the fibula, from the head of the tibia and from the interosseous membrane; its fibres are inserted in a penniform manner into three tendons which commence upon the anterior border of the muscle; these pass through the annular ligament, and on the dorsum of the foot the inner tendon divides into two, thus making four tendons destined to all the toes except the great toe. The tendons of the extensor brevis are inserted into the outer side of each of these tendons, with the exception of the one to the little toe, and the expanded tendon thus formed divides, as in the hand, into three slips, the central one of which is inserted into the second phalanx, and the two lateral into the sides of the last phalanx.

The EXTENSOR BREVIS DIGITORUM, covered in by the tendons of the extensor longus and peroneus tertius muscles, lies upon the dorsum of the foot, and consists of a small flat belly arising from the outside of the os calcis, and from which emanate four tendons; the first crosses the dorsalis pedis artery, and is inserted into the first phalanx of the great toe, the other three terminate by blending with the outer side of the tendons of the extensor longus muscle.

The PERONEUS TERTIUS MUSCLE is in reality a part of the extensor longus; it arises from the lower third of the fibula as a thin layer of fibres, which terminate in a round tendon that passes under the annular ligament in the same sheath with the extensor longus, and is inserted into the base of the metatarsal bone of the fifth toe. It is not unfrequently wanting.

Upon separating the tibialis anticus and the extensor longus muscles, the *anterior tibial artery* will be found, lying deep in the upper part of the leg, but more superficial lower down. After its division from the popliteal artery, it pursues a short course posteriorly, to the space between the fibula and tibia, and becomes anterior by passing through the interval left at the upper part of the interosseous membrane, upon which it descends, in company with the anterior tibial nerve and two large venæ comites; in the upper part of its course it lies between the tibialis anticus and the extensor longus muscles; lower down it lies between the tibialis anticus tendon and that of the extensor proprius pollicis, passes under the annular ligament, and on the back of the foot becomes the *dorsalis pedis artery*. At its upper part it sends a *recurrent* branch upward, which perforates the head of the tibialis anticus muscle, and anastomoses with the inferior articular branches of the popliteal; in its course between the muscles it gives off numerous *muscular* twigs, and at the ankle-joint an *external* and *internal malleolar* branch; these supply the parts about the joint, anastomosing with the calcanear branches of the posterior tibial and peroneal arteries.

The *anterior peroneal artery*, coming from the back of the leg, passes through an aperture in the lower part of the interosseous membrane, and is distributed in front of the fibula to the dorsum and outer part of the foot, anastomosing with the malleolar branches of the anterior tibial.

The *anterior tibial nerve*, one of the divisions of the peroneal nerve, having passed through the opening in the upper part of the interosseous membrane, will be found emerging from under the belly of the extensor longus muscle, to accompany the anterior tibial artery which lies upon its outer side; on the dorsum of the foot it is distributed as has been described.

The *musculo-cutaneous nerve*, the other division of the peroneal nerve, after curving round the head of the fibula and passing beneath the fascia of the leg, lies between the extensor longus and the peronei muscles; it then pierces the fascia, becomes subcutaneous, and is distributed as has been described.

The *dorsalis pedis artery* lies upon the outer side of the tendon of the extensor proprius pollicis; it is accompanied by the anterior tibial nerve; it distributes some branches to the tarsus, called *tarsal*, and then forms an arch over the bases of the metatarsal bones; this arch gives off three *interosseous* branches which, at the commissures of the third, fourth, and fifth toes, divide into *digital* branches, to supply the sides of the toes. At each end of the interosseous spaces these arteries are joined by perforating branches from the sole of the foot. At the base of the first interosseous space, the dorsalis pedis penetrates to the sole of the foot, to unite, under the name of the *communicating artery*, with the termination of the plantar arch; before disappearing it gives off the *dorsalis hallucis* branch, which passes forward and is distributed to the great toe and the inner side of the next toe.

The DORSAL INTEROSSEOUS MUSCLES arise by two heads from the sides of the bases of adjoining metatarsal bones; they are inserted into the bases of the first phalanges of the toes, and into the expansion of the extensor tendons; the first interosseous muscle being inserted into the inside of the second toe, and the three others into the outsides of the second, third, and fourth toes.

DISSECTION XII.

BACK OF THE LEG.

The dissection of the back of the leg is to be commenced by an incision from the popliteal space to the heel. The integument is to be reflected with care, as there are several superficial structures to be examined.

The *external saphena vein*, commencing on the outer border of the foot, ascends along the outside of the tendo Achillis, upon the belly of the gastrocnemius muscle and between its two heads, to the popliteal space, where it enters the popliteal vein.

The *external saphena nerve*, formed from the communicans peronei and poplitei, which unite at a variable distance down the leg, or occasionally, in the case of their non-union, being one of these nerves itself, lies at the side of the external saphena vein along the outer border of the tendo Achillis; it curves around the external malleolus and is distributed to the outer border of the foot and little toe.

The muscles of the leg are arranged in two layers, superficial and deep; the superficial layer constitutes the "calf of the leg."

The GASTROCNEMIUS MUSCLE, the first muscle of the superficial layer, arises by two heads, of which the inner is the larger, from the surface of bone above each condyle of the femur; these heads converge and form the lower boundary of the popliteal space, and are situated inside the tendons forming the ham-strings; they are each supplied with an arterial branch, the *sural*, from the popliteal artery. The large muscular belly of the gastrocnemius terminates in a brilliant aponeurosis, finally converted into a large, round tendon, called the *tendo Achillis*, which is inserted into the lower part of the posterior surface of the os calcis; this tendon, for some distance above its insertion, is common to this muscle and to the soleus which lies beneath it.

The small rounded tendon of the plantaris muscle, running along the inside of the tendo Achillis, to which it is more or less adherent, should be sought for and isolated, that it may not be cut across in dividing the gastrocnemius.

The gastrocnemius should be separated along its borders from the soleus, in order to determine the line of division between them; it may then be cut across just below the union of its heads, and the two halves reflected.

In dissecting up the heads of the gastrocnemius, the strong character of their osseous attachment will be seen. A synovial bursa is sometimes found between the muscle and the condyle on one or both sides, and a sesamoid bone is occasionally developed in the head attached to the outer condyle.

The PLANTARIS MUSCLE arises from the outer condyle of the femur in common with the under part of the outer head of the gastrocnemius; it forms a short belly, from two to four inches in length, terminating in a long, slender tendon which crosses obliquely between the gastrocnemius and soleus muscles, and passing downward along the inner side of the tendo Achillis, is inserted by the side of that tendon into the calcaneum. It is not always easy to separate it from the tendo Achillis at its lower part.

The SOLEUS MUSCLE arises from the head and upper half of the fibula, and from the middle of the shaft of the tibia; it forms an elliptical-shaped belly, and joins the tendo Achillis some distance below its commencement. A synovial bursa is placed between the tendo Achillis and the calcaneum above the point of its insertion, and a considerable interspace, filled with fat and cellular tissue, exists between it and the layer of muscles beneath.

The tendo Achillis is to be divided, and the soleus and plantaris entirely removed; in doing this, it must be remembered, that the deep vessels and nerves, covered in by a fascia, lie between these muscles and those beneath.

A stout layer of fascia covers in the deep muscles and vessels; it extends from the popliteal space to the ankle-joint, and is attached to the fibula on one side, and the tibia on the other.

The popliteal artery will be found dividing into the anterior and posterior tibial arteries, and the course of the *anterior tibial artery*, to the point where it perforates the interosseous membrane, will now be seen.

The *posterior tibial artery* descends the leg on the side of the tibia, and, at the lower part of its course, becomes comparatively superficial; it runs along the tendo Achillis, to the concavity formed by the internal malleolus and os calcis, where it divides into the plantar arteries, which will be seen in the dissection of the sole of the foot. It gives off in its course numerous muscular branches, some of which were divided in removing the superficial layer of muscles; others will be seen going to the deep layer; a *nutrient artery*, to the tibia, may sometimes be found, if the subject is well injected. The principal branch of the posterior tibial artery is the peroneal.

The *peroneal artery* arises some distance from the commencement of the posterior tibial, and passes downward along the inner border of the fibula; in the lower third of the leg it divides into two branches: one of them, the *anterior*, perforates the interosseous membrane, and is distributed in front of the external malleolus; the *posterior* division continues downward to the outside of the os calcis, where it breaks up into *external calcanear branches*. The peroneal and posterior tibial arteries present frequent variations of distribution and size; one or the other is occasionally absent, and they sometimes communicate with each other by means of a short, though large, transverse anastomosis.

The *posterior tibial nerve* accompanies the posterior tibial artery, lying first upon its inside, and then upon its outside. At the inner malleolus, it divides into two branches, the internal and external plantar. In its course, it gives off muscular branches, one of which accompanies the peroneal artery; at the heel, it gives off the *plantar cutaneous branches*, distributed to the integument of the side of the heel.

The POPLITEUS is a short muscle situated just below the knee-joint; arising from the outer side of the external condyle, it passes obliquely downward, to be inserted into the surface of the tibia, above the oblique line, known as the popliteal.

The FLEXOR LONGUS DIGITORUM PEDIS arises from the surface of the tibia, below the popliteus muscle, and passing behind the internal malleolus, through a sheath in the internal annular ligament, between the tendons of the tibialis posticus and the flexor longus pollicis, enters the sole of the foot, and divides into four tendons, inserted into the last phalanges of all the toes, except the great toe, as will be seen hereafter.

The FLEXOR LONGUS POLLICIS PEDIS lies upon the outer side of the leg, and arises from the lower two thirds of the fibula, and from the interosseous membrane; its tendon curves around the internal malleolus, passes through a sheath of the internal annular ligament, below the tubercle of the os calcis, and, entering the sole of the foot, is inserted into the last phalanx of the great toe. The posterior tibial nerve lies along the inner side of this muscle, and the peroneal vessels are in part concealed by it.

The TIBIALIS POSTICUS MUSCLE occupies a position between the two bones of the leg, and between the two muscles last described. It arises from nearly the whole length of both tibia and fibula, and from the interosseous membrane; its origin superiorly forms, as it were, two heads, between which passes the anterior tibial artery; its tendon curves around the internal malleolus, passing through the sheath in the internal annular ligament which is nearest the malleolus, and enters the sole of the foot, to be inserted into the scaphoid and external cuneiform bones, and into the base of the first metatarsal bone.

Along the outer border of the leg, and upon the fibula, will be found two muscles, the peroneus longus and peroneus brevis.

The PERONEUS LONGUS MUSCLE arises from the head and upper part of the surface of the fibula, and terminating in a long tendon, curves around the external malleolus, and passes through the sheath of the external annular ligament; entering the sole of the foot, it crosses obliquely forward to be inserted into the base of the metatarsal bone of the great toe. The musculo-cutaneous nerve lies between this muscle and the extensor-longus digitorum.

The PERONEUS BREVIS lies beneath the preceding muscle, and arises from the lower half of the outer surface of the fibula; its fibres are inserted in a penniform manner into a tendon which passes beneath the external annular ligament, with the peroneus longus, and is inserted into the base of the metatarsal bone of the fifth toe.

The INTERNAL ANNULAR LIGAMENT confines the tendons which curve around the inner ankle from the back of the leg; it stretches across from the tip of the internal malleolus to the side of the os calcis, and contains three compartments, lined by synovial membrane, which transmit the tendons of the tibialis posticus, flexor longus digitorum pedis, and flexor longus pollicis, in the order in which they have been mentioned, that of the tibialis posticus being nearest the malleolus.

The EXTERNAL ANNULAR LIGAMENT stretches across from the tip of the external malleolus to the side of the os calcis, and contains a single compartment lined with synovial membrane, through which pass the tendons of the peroneus longus and brevis muscles.

DISSECTION XIII.

SOLE OF THE FOOT.

The density of the cuticle, the thickness of the fat and areolar tissue superficially, and the number and smallness of the muscles, with the amount of aponeurotic structure which belongs to the region, render the dissection of the sole of the foot a slow and tedious process. A block should be put under the instep, and as the foot, by its own weight, offers no impediment to constant unsteadiness, it should be fixed to the table by hooks. An incision is to be made down the middle of the sole, from the heel to the commissure of the toes; this should penetrate to the plantar fascia; the integument, with the fat, should then be removed, and, as a preliminary step, this fascia should be cleanly dissected, and the deposits of fat removed from its interstices, carefully looking out for the nerves which nearly resemble its fibres in color.

The *plantar fascia* is a thick aponeurosis which expands over the whole sole of the foot, lying between the muscles and the adipose tissue beneath the integument; it is thicker in the centre than at the sides of the foot, and dividing into slips anteriorly, is attached to the base of the first phalanx of each toe by lateral processes, between which pass the flexor tendons; these are crossed by some transverse bands of fibres which form the rudimentary web of the commissures of the toes; between the slips, the nerves destined to the toes, and the lumbricales muscles will be found.

The plantar fascia may be removed by dividing it transversely and dissecting it up from the subjacent attachments; this will expose the muscles beneath. In direct relation with the central portion of the plantar fascia is the flexor brevis digitorum.

The FLEXOR BREVIS DIGITORUM PEDIS arises from the under side of the os calcis and from the plantar fascia; it divides into four tendons, which are inserted into the bases of the second phalanges of the outer four toes; they are each perforated to permit the passage through them of the tendons of the long flexor, which is inserted into the last phalanx.

The *sheaths* of the flexor tendons consist of transverse fibrous bands which hold the tendons down upon the phalanges; though similar to those of the fingers, they are not so distinct nor so well developed.

The ABDUCTOR POLLICIS PEDIS lies upon the side of the great toe; it arises by two heads, which are not, however, well-marked divisions; one springs from the internal annular ligament which covers in the vessels and tendons below the inner malleolus, the other from the inner tuberosity of the os calcis. Between these heads pass the plantar vessels, nerves, and tendons. It is inserted, after uniting with the tendon of the flexor brevis pollicis, into the base of the first phalanx of the great toe.

The ABDUCTOR MINIMI DIGITI PEDIS is placed along the outer edge of the foot; it arises from the external surface of the os calcis, and from the plantar fascia which covers it, and is inserted into the base of the first phalanx of the little toe.

The flexor brevis digitorum is separated from the muscles on each side of it by strong intermuscular septa; the inner of these is perforated by the internal plantar nerve and by the tendon of the flexor longus pollicis; the outer by a nerve and artery destined to the little toe.

The flexor brevis digitorum and abductor minimi digiti should be divided, and their two ends reflected; the tendons of the long flexor, with the plantar vessels and nerves, will then be exposed, and should be cleared from the superfluous areolar tissue which invests them.

The posterior tibial artery, on entering the sole of the foot, divides into the internal and external plantar arteries.

The *internal plantar artery* passes between the abductor pollicis and flexor brevis digitorum to the great toe; it gives off small muscular branches, and terminates in supplying the sides of the great toe; it anastomoses with the digital branches of the external plantar artery, and the communicating branch of the dorsalis pedis.

The *external plantar artery* is larger than the internal; it passes obliquely beneath the flexor brevis digitorum to the base of the fifth metatarsal bone, and then curves transversely across the foot, dipping beneath the deeper muscles, to the first interosseous space, which it penetrates to anastomose with the dorsalis pedis artery.

The nerves of the sole of the foot are derived from the posterior tibial nerve, which, at the inner malleolus, divides into external and internal plantar branches.

The *internal plantar nerve* is the larger of the two, and runs along the edge of the abductor pollicis muscle, and divides opposite the bases of the metatarsal bones, into four digital branches which supply, by bifurcating, both sides of the first, second, and third toes, and the inside of the fourth toe; it also gives off small muscular branches.

The *external plantar nerve* accompanies the external plantar artery as far as the fifth metatarsal space, where it sends off a large muscular branch, and continues on in two divisions, which supply the outer edge of the foot, the little toe, and the outside of the next, which is unsupplied by the internal plantar.

In the second layer of muscles are found the tendons of the flexor longus digitorum and flexor longus pollicis muscles, the bellies of which have been already dissected upon the back of the leg (p. 243). The tendon of the flexor longus digitorum passes to the middle of the sole of the foot, and is there joined by a tendinous process from the flexor longus pollicis, and by the musculus accessorius; it then divides into four tendons for the outer four toes; these perforate the tendons of the flexor brevis, and are held down to the phalanges by ligamenta brevia, like those of the tendons of the fingers (p. 154). Their insertion and that of the flexor longus pollicis have previously been given.

The LUMBRICALES MUSCLES are four small muscles arising from the tendons of the long flexor, and inserted into the expansions of the extensor tendons, and the inner sides of the first phalanges of all the toes except the great toe. They are subject to variations.

The MUSCULUS ACCESSORIUS, or MASSA CARNEA JACOBI SYLVII, arises by two heads, one from the concave, inner side of the os calcis, and the other from its under surface, and is inserted into the tendon of the flexor longus digitorum, just as it breaks up into its four digital divisions.

The tendons of the flexor longus must be divided, and, with their accessory muscles, reflected.

The FLEXOR BREVIS POLLICIS PEDIS arises from the cuboid and external cuneiform bones; it lies upon the metatarsal bone of the great toe, and is inserted by two heads into the base of its first phalanx; the outer head joins with the tendon of the abductor pollicis, and the inner head with that of the adductor pollicis; a sesamoid bone is usually found in each of these heads; the tendon of the flexor longus pollicis passes between them.

The ADDUCTOR POLLICIS PEDIS is placed obliquely in the sole of the foot; it arises from the cuboid bone, the bases of the third and fourth metatarsal bones, and from the sheath of the tendon of the peroneus longus muscle; it forms a short belly, and is inserted, with the inner head of the flexor brevis pollicis, into the base of the first phalanx of the great toe. A sesamoid bone is usually found in the insertion of its tendon.

The FLEXOR BREVIS MINIMI DIGITI PEDIS arises from the base of the metatarsal bone of the fifth toe and the sheath of the tendon of the peroneus longus, and is inserted into the base of the first phalanx of the little toe.

The TRANSVERSALIS PEDIS is a thin layer of muscular slips, lying transversely across the anterior portions of the metatarsal bones; it arises from the heads of the metatarsal bones of the four lesser toes, and is inserted into the tendon of the adductor pollicis; sometimes there is but one slip; there are rarely so many as four.

The adductor pollicis and flexor brevis minimi digiti are to be divided near their origins, and turned forward on to the toes; this will expose the plantar arch.

The *plantar arch* is the continuation of the external plantar artery from the base of the fifth metatarsal bone to its anastomosis with the communicating branch of the dorsalis pedis; it lies upon the interosseous muscles, and gives off four *digital* branches to the toes; the first goes to the outer side of the little toe; the others, dividing at the commissures, supply the contiguous sides of the outer three toes and the outer side of the second. At each end of the interosseous spaces, the digital arteries send small branches called *posterior* and *anterior perforating*, to the interosseous branches of the dorsalis pedis artery. The communicating branch of the dorsalis pedis enters the sole of the foot, and, uniting with the internal plantar artery and the termination of the plantar arch, gives off the branch which supplies both sides of the great toe.

The PLANTAR INTEROSSEOUS MUSCLES are three in number; they arise from the under sides of the metatarsal bones, and are inserted into the inner sides of the bases of the first phalanges of the outer three toes, and into the expansions of the extensor tendons.

The tendon of the tibialis posticus (p. 244) may now be traced to its termination; it passes forward over the articulation of the astragalus and scaphoid bones to be inserted into the latter, into the base of the first metatarsal bone, and also into the external cuneiform bone; a sesamoid bone is usually found in its tendon.

The tendon of the peroneus longus (p. 244) passes obliquely across the foot to the base of the metatarsal bone of the great toe, into which it is inserted; as it turns round the cuboid bone its tendon becomes thickened and fibrocartilaginous, and sometimes contains a sesamoid bone; it is enveloped with a sheath formed by the ligaments of the tarsal bones, and lined by a synovial membrane.

DISSECTION XIV.

LIGAMENTS OF THE PELVIS AND LOWER EXTREMITY.

The pelvis is connected with the vertebral column by ligaments similar to those uniting one vertebra to another, with the addition of two special ligaments, the lumbo-sacral and the ilio-lumbar.

The *lumbo-sacral ligament* is a stout, triangular bundle of fibres springing from the tip of the transverse process of the last lumbar vertebra, and expanding, fan-shaped, to be inserted into the posterior part of the upper border of the sacrum.

The *ilio-lumbar ligament* extends between the tip of the transverse process of the last lumbar vertebra and the crest of the ilium, just above the sacro-iliac articulation.

The sacrum and the coccyx are united by an *anterior* and *posterior common ligament*, and by intervening *fibro-cartilaginous disks;* in adults, the bones are, however, usually co-ossified.

The sacrum and ilium are united by cartilage, and by anterior and posterior sacro-iliac ligaments, at the sacro-iliac synchondrosis, and by the sacro-ischiatic ligaments inferiorly.

The *anterior sacro-iliac ligament* is a transverse band of fibres covering the anterior aspect of the articulation.

The *posterior sacro-iliac ligament* is composed of stout bundles of fibres passing between the first two bones of the sacrum and the rough surface at the posterior border of the ilium.

The *greater sacro-ischiatic ligament* passes from the side of the sacrum and coccyx to the tuberosity of the ischium.

The *lesser sacro-ischiatic ligament* arises from the side of the sacrum and coccyx, and is inserted into the spine of the ischium. These two ligaments convert the space between the sacrum and os innominatum into two apertures, called the greater and lesser sacro-ischiatic foramina, through which issue the nerves, arteries, and muscles from the interior of the pelvis.

The union of the pubic bones in front is called the *symphysis pubes*.

The *anterior pubic ligament* consists of horizontal and oblique fibres, interlacing in front of the symphysis. The periosteum constitutes the *posterior pubic ligament*.

The *superior pubic ligament* covers the surface of the bones superiorly, and blends with the tendinous insertions of the abdominal muscles.

The *sub-pubic ligament* is a fibrous arch attached to the bones of the pubes inferiorly, and losing itself at each side on the rami of the ischia.

The *inter-articular fibro-cartilage*, seen on opening the symphysis, is composed of concentric layers of fibres firmly attached to the opposed surfaces of the bones, and projecting a little beyond their borders. At the posterior part, a cavity, containing a fluid like the synovial, and of variable size, will be found separating the cartilage into two lateral halves; this cavity is said to increase in size during pregnancy.

The *obturator ligament*, or *membrane*, is a tendino-fibrous expansion, which stretches across the obturator foramen, and closes it in its entire extent, except at the upper part, where the obturator artery and nerve pass out of the pelvis.

The COXO-FEMORAL ARTICULATION, or HIP-JOINT, in which the head of the femur is received into the cotyloid cavity of the os innominatum, is maintained by a capsular and an inter-articular ligament.

The *capsular ligament* extends from the circumference of the acetabulum to the anterior inter-trochanteric line of the femur in front and to the neck of the bone posteriorly. It is strengthened anteriorly by a band of fibres, called the *ilio-femoral ligament*, which passes from the anterior inferior spinous process of the ilium to the inter-trochanteric line. The anterior portion of the capsular ligament is of great strength, and that part of it arising from the anterior inferior spinous process of the ilium and inserted into the inter-trochanteric line, including what is known as the ilio-femoral ligament, divides inferiorly into two bands, separated by an interval, their disposition resembling an inverted letter Y. This arrangement of the fibres and its influence in maintaining, as well as causing the position of the head of the femur, characteristic of its various forms of dislocation, and in the reduction of these by manipulation, have been particularly described by Dr. H. J. Bigelow.

The capsular ligament is to be divided transversely, and the head of the femur dislocated from its socket, to see the inter-articular ligament. Or, better still, a circular portion of bone may be removed on the inside of the os innominatum, by the gouge, from the space between the obturator foramen and the greater sacro-ischiatic notch, cutting out the floor of the acetabulum; this shows the ligament from within.

The *inter-articular ligament*, or *ligamentum teres*, extends from the triangular depression in the head of the femur to the borders of the notch in the acetabulum, where it blends with the fibres of the transverse ligament.

The *cotyloid ligament* is a fibro-cartilaginous band attached to the margin of the acetabulum, the cavity of which is surrounded and deepened by it.

The *transverse ligament* is a band of fibres, continuous with those of the cotyloid ligament, which extends across the notch of the acetabulum, protecting the vessels which pass beneath it to the inter-articular ligament and head of the femur.

The head of the femur is not entirely coated with cartilage, a portion of it around the depression in its centre being divested of that covering. In the cotyloid cavity there is also a depression which has no cartilage; this is occupied by a mass of fat, sometimes called the *synovial gland of Havers*.

The TIBIO-FEMORAL ARTICULATION, or KNEE-JOINT, is invested by a capsule, and by anterior, posterior, external and internal, lateral ligaments.

The *capsule* is a fibrous membrane which surrounds the heads of the bones, and fills the intervals of the stronger special ligaments. It is connected with the patella, femur, tibia and the inter-articular cartilages, and is lined internally by the synovial membrane.

The *external lateral ligament* is a rounded bundle of fibres, extending between the external condyle and the head of the fibula; a second bundle is sometimes found posterior to this, and called the *short external lateral ligament.*

The *internal lateral ligament* is attached to the internal condyle above, and the inner tuberosity of the tibia below; its limits are not distinctly marked; the inferior internal articular artery, and the tendon of the semi-membranosus, pass beneath it.

The *anterior ligament*, or *ligamentum patellæ*, is attached to the lower border of the patella above, and the tubercle of the tibia below; a bursa will be found beneath its insertion into the tubercle.

The *posterior ligament*, or *ligamentum posticum Winslowii*, is formed chiefly by fibres of the tendon of the semi-membranosus muscles, which pass across the joint to the outer condyle; a deeper set of fibres is continuous with the general capsule.

The knee-joint is to be opened by an incision along each side and across the front, above the patella; the patella and its ligament are to be turned downward; this exposes the synovial membrane and inter-articular ligaments.

The *synovial membrane*, coextensive with the capsule, covers the cartilages of the bones, and all the inter-articular structures; it forms a pouch on each side of the patella, extending above and below it a distance sometimes as great as two inches; posteriorly it sends a pouch between the head of the popliteus muscle and the tibia. A fold of the synovial membrane, called the *mucous ligament*, extends from between the condyles to the fat below the patella; this is continued outward to the sides of the patella, under the name of the *alar ligaments*.

The *crucial ligaments* are two strong bands, the *anterior* of which, arising from the depression in front of the spine of the tibia, goes to the inner surface of the external condyle, and the *posterior* from the depression behind the spine to the inner surface of the internal condyle; the internal ligament is larger than the external. The respective insertions of the crucial ligaments are difficult to remember; the initial letters of the words, anterior external, posterior internal, manufacture the word AEPI, from which the memory may derive assistance.

The *semi-lunar fibro-cartilages* are two crescentic plates of cartilage placed upon the margins of the head of the tibia, and attached externally to the capsule of the joint; they are thick along their convex borders and thin at their concave, being hollowed out to receive the condyles of the femur; they are connected anteriorly with the front of the tibia, and posteriorly with its spine. The internal cartilage forms the segment of a larger circle than the external, and is more ovoidal in shape.

The *transverse ligament* unites these cartilages anteriorly; sometimes it hardly exists.

The tibia and fibula are articulated together at their two extremities, and connected between their shafts by an interosseous membrane.

The union between the bones superiorly is effected by *anterior* and *posterior* bands from the tuberosity of the tibia to the head of the fibula. Inferiorly, an *anterior* band crosses in front from the fibula to the tibia, and a *posterior* one is similarly disposed behind the ankle-joint; an *inferior interosseous ligament* closes the space between the ends of the two bones below, and may be seen by forcibly tearing them apart.

The *interosseous membrane* is attached to the contiguous sides of the shafts of the tibia and fibula, and separates the muscles of the back from those of the front of the leg; its fibres are directed downward and outward, and are crossed by a few passing in the opposite direction; they are deficient above, at the point where the anterior tibial artery enters, and below, where the anterior peroneal artery comes forward from the back of the leg, to be distributed upon the tarsus.

The TIBIO-TARSAL ARTICULATION, or ANKLE-JOINT, is formed by the fibula, the tibia, and the astragalus, and is maintained by four ligaments, anterior, posterior, external, and internal.

The *anterior ligament* is a thin membranous layer, extending from the front of the tibia to the upper part of the astragalus, and continuous by its sides with the lateral ligaments.

The *internal lateral*, or *deltoid ligament*, expands, fan-shaped, from the tip of the malleolus, to be inserted into the inner side of the astragalus, os calcis, and scaphoid bone. The ligament is covered in by the internal annular ligament (p. 244), and the tendons which pass through it.

The *external lateral ligament* consists of three strong fasciculi; the *anterior* of which passes from the anterior border of the malleolus to the surface of the astragalus in front of it; the *middle* descends from the tip of the malleolus to the side of the os calcis, and the *posterior* from the posterior border of the malleolus, horizontally backward, to the posterior surface of the astragalus.

The tarsal bones of the foot are united by dorsal, plantar, and interosseous ligaments.

The *dorsal ligaments* unite each tarsal bone with those contiguous to it; several strong bands pass forward from the deep fossa between the astragalus and os calcis, to the scaphoid and cuboid bones, under the name of *calcaneo-cuboid* and *astragalo-scaphoid ligaments*.

The *plantar ligaments* also consist of short bands uniting the contiguous bones, with the addition, however, of two others of large size and great strength. The *inferior calcaneo-scaphoid ligament* unites the os calcis and the scaphoid bone, forming part of the cavity which receives the rounded head of the astragalus; the tendon of the tibialis posticus crosses this ligament. The *long calcaneo-cuboid ligament* passes from the under surface of the os calcis to the rough ridge on the under part of the cuboid bone; some of its fibres continue over the tendon of the peroneus longus, forming a sheath for it, and are inserted into the bases of the third and fourth metatarsal bones; these last-named fibres are sometimes called the *ligamentum longum plantæ*, and the shorter ones the *ligamentum breve plantæ*.

The *interosseous ligaments* are five in number, and are strong bands intervening between the contiguous surfaces of the adjoining tarsal bones. The *calcaneo-astragaloid* lies in the fossa between the os calcis and astragalus, and can only be seen by a longitudinal section of the two bones, when it will be found as a short, stout band, attached to a depression in each bone. Another interosseous ligament exists between the articulating surfaces of the cuboid and scaphoid bones, and three others between the three cuneiform and the cuboid bones.

The bases of the metatarsal bones are united by *dorsal*, *plantar*, and *interosseous ligaments;* the interosseous ligaments are to be demonstrated by tearing the bones apart, when the dissection is completed. These ligaments are only found between the outer four metatarsal bones, that of the great toe not being united with the others; the second and third metatarsal bones are also firmly connected with the external and internal cuneiform bones. The heads of the metatarsal bones are connected inferiorly by *transverse metatarsal ligaments*.

The *metatarso-phalangeal articulations* are maintained by two *lateral* ligaments and an *inferior* ligament, the expansion of the extensor tendon supplying the place of a superior one.

The phalanges are united to each other by two *lateral* ligaments and an inferior ligament, and by the expansion of the extensor tendon superiorly. The joints between the phalanges are sometimes very indistinct, and occasionally co-ossified. The ligamentous arrangements and the insertions of the tendons are precisely the same as in the fingers, and are much more satisfactorily studied in the hand, the description of which is given at p. 160.

PECULIARITIES

IN THE

ANATOMY OF THE FŒTUS.

Upon the fœtus may be demonstrated almost all the anatomy of the adult human system; the points in which it differs from that of adults will alone be noticed.

The full-grown fœtus weighs about seven pounds, and is seventeen inches in length, or thereabout; the umbilicus is situated from a quarter to half an inch below the middle of the body. The external genital organs are largely developed, especially the labia minora in the female, which project beyond the labia majora; in the male, the prepuce is adherent to the glans penis, constituting a state of phymosis. The muscles are pale, and of a softer texture than when more developed, and are covered in with a dry and granular fat, easily detached from them.

The umbilical cord is composed of two umbilical arteries, and an umbilical vein, imbedded in a soft, semi-transparent substance, called the *Whartonian gelatine*. Within the abdomen, the arteries are called the hypogastric.

The abdomen is to be opened in such a way as to avoid injuring the vessels which diverge upward and downward from the umbilicus.

The HYPOGASTRIC ARTERY is the internal iliac artery of adult life; it is larger than the external iliac, and may be traced from the common iliac artery upward, along the side of the urinary bladder and urachus, to the anterior wall of the abdomen, upon which it ascends to the umbilicus, where the two arteries come together, escape from the abdomen with the umbilical vein, and coil around it in the umbilical cord, until they reach the placenta. After the cessation of the placental circulation, this artery becomes impervious from the side of the bladder upward, and is converted into a fibrous cord (p. 208); an inch or so of its commencement always remains pervious, and gives origin to the superior vesical artery.

The UMBILICAL VEIN runs from the umbilicus along the free margin of the suspensory ligament of the liver to the transverse fissure of that organ; it there divides into three branches, one of which is distributed to the left lobe, one to the right lobe, larger than the preceding, and which is joined by the vena portæ, and a third branch, smaller than either of the others, the *ductus venosus*, which terminates in the left hepatic vein. After the cessation of the placental circulation, the umbilical vein and ductus venosus become converted into a fibrous cord, the former becoming the round ligament of the liver.

The LIVER occupies the whole upper part of the abdomen; it is of a dark, mahogany color, and its lobes are nearly equal in size. After birth, it rapidly diminishes in bulk; and, at the age of five or six years, attains the proportions maintained during the rest of life.

The KIDNEYS present a lobulated appearance, and are relatively larger than in the adult. The *supra-renal capsules* are also of large size.

The BLADDER is long and conical, and is connected superiorly with the umbilicus by a fibrous cord called the *urachus*. This is an obliterated tube which during the early part of intra-uterine life connected the bladder with one of the fœtal membranes called the allantois.

The INTESTINES are small in calibre, and contain a dark green substance called *meconium*. The small intestines are devoid of valvulæ conniventes, or these are but imperfectly developed. The appendix of the cæcum is long and of large size, and seems like a tapering continuation of the cæcum itself.

The TESTES, in the early part of fœtal life, are situated in the lumbar region, behind the peritoneum. About the fifth month they begin to descend to the scrotum, but not unfrequently they may be found delayed in some part of their course at the time of birth. Connected with the lower end of the testicle and epididymis is a band composed of areolar and muscular tissue, called the *gubernaculum testis*, which guides and assists their gradual descent; it extends through the inguinal canal, and is attached to the front of the pubes and the bottom of the scrotum. When the testicle is about to enter the internal abdominal ring, a small pouch of peritoneum will be found protruding at that point, and into this the testicle projects from behind, supported by a duplicature, or suspensory fold, of the peritoneum, called the *mesorchium;* this pouch, under the name of *processus vaginalis peritonei,* precedes the testicle in its course through the inguinal canal, and enters the scrotum in advance of the gland. The neck of this pouch, by which it is connected with the general peritoneal cavity, becomes gradually obliterated after birth, while the pouch itself remains as an independent serous sac, under the name of tunica vaginalis testis.

The *ovaries* in the female are likewise placed in the lumbar region, and gradually descend to the pelvis. A pouch of peritoneum, analogous to the processus vaginalis of the male, accompanies the round ligament of the uterus for a short distance into the inguinal canal, and is called the canal of Nuck.

On opening the thorax there is found in the anterior mediastinum, and extending upward upon the trachea into the neck, a narrow, elongated body, lobulated, pinkish in color, and soft in texture; this is the THYMUS GLAND. It consists of two lateral lobes enveloped in an areolar capsule. A central cavity, containing a milky fluid, exists in each lobe; it has no duct or outlet. The thymus gland increases in size for about two years; it then dwindles, and becoming converted into a fatty mass, at the age of puberty has nearly disappeared.

The LUNGS are small, and lie packed in the posterior part of the thorax. Previous to respiration they consist of a dense, gland-like substance. Subsequently to respiration they are of a pinkish, spongy structure, expanding and completely filling the pleural cavity.

The HEART is well developed in point of size, but the septum between the auricles is incomplete, being perforated by a large opening called the *foramen ovale.* A communication will also be found between the left pulmonary artery and the aorta, just beyond the origin of the brachio-cephalic vessels, by a short trunk called the *ductus arteriosus;* this degenerates into a fibrous cord shortly after birth, and the foramen ovale usually becomes obliterated. The Eustachian valve is of large size, and seems to be a continuation upward of the anterior wall of the inferior vena cava toward the foramen ovale.

The FŒTAL CIRCULATION presents peculiarities. The

oxygenized blood from the placenta is brought by the umbilical vein to the inferior vena cava, where it of course mixes with the impure blood from the lower extremities; entering the right auricle, its current is directed by the large Eustachian valve toward the foramen ovale, through which it passes into the left auricle. The blood from the head and upper extremities, which is returned by the superior vena cava as venous blood, enters the right auricle, and passes by the auriculo-ventricular opening into the right ventricle; the crossing of these two currents in the right auricle is permitted by the construction of the Eustachian valve, though, to a certain extent, the two streams must intermingle.

We thus see that the blood brought to the right auricle has two sources of exit, viz: the foramen ovale and the auriculo-ventricular orifice. By following the blood from the cavities into which these orifices open, as if it had started originally from these points, we shall at once comprehend the fœtal circulation. Thus, that from the right ventricle must enter the pulmonary artery which goes to the lungs; but the lungs are solid and impervious, and nature has therefore provided an exit for it by the ductus arteriosus, through which it reaches the aorta at a point beyond the origin of the brachio-cephalic branches, and is carried to the body and lower extremities. Then, again, the blood which is in the left auricle can go only to the left ventricle, and from there it must go to the aorta. This portion of the blood being that from the placenta, and therefore the most richly oxygenized, is able to enter the brachio-cephalic trunks, which the other current could not do, and supplies the important parts to which they are distributed. Beyond the origin of the brachio-cephalic trunks that portion of the blood which has not entered those vessels, joins with the current from the ductus arteriosus, helps to supply the body and lower extremities, and is returned to the placenta by the hypogastric arteries.

IMPORTANT ANATOMICAL LANDMARKS AND POINTS,

CAPABLE OF BEING STUDIED WITHOUT DISSECTION, OR UPON THE LIVING SUBJECT.

CRANIUM.

THE position of the *lateral sinus* is indicated by a line nearly horizontal, drawn from the occipital protuberance, which may be felt at the back of the head, to the base of the mastoid process of the temporal bone.

The position of the *longitudinal sinus* is indicated by a line drawn, over the vertex, from the root of the nose to the occipital protuberance.

The *middle meningeal artery* follows a course upward from the anterior inferior angle of the parietal bone, or from a point about one and a half inches behind the external angular process of the frontal bone.

It is not usual to apply the trephine over the region traversed by the lower part of the meningeal artery, or over those corresponding to either of the above named sinuses.

FACE.

The *supra-orbital foramen*, from which issues the supra-orbital branch of the fifth nerve, is situated a little inside of the union of the inner with the outer two-thirds of the upper margin of the orbit. Its position may also be determined by the pulsations of the supra-orbital branch of the ophthalmic artery, issuing from it, and which, though the vessel is of small size, may still be detected.

The *infra-orbital foramen*, from which issues the infra-orbital branch of the fifth nerve, is situated just above the canine fossa, one-fourth of an inch below the lower edge of the orbit. A vertical line dropped from this foramen would fall upon the first molar tooth.

The *lachrymal canals*, superior and inferior, are situated at the inner angle of the eyelids, in their free margin. Their orifices, or puncta, present slight prominences, are directed inward towards the globe of the eye, and it is necessary that the lid should be slightly everted to see them. The canals are at first directed vertically, the superior from below upward, and the inferior from above downward; they then speedily bend at a right angle and continue inward to the lachrymal sac, which they usually enter by separate orifices. The length of the canals is three or four lines, and the lower is a little shorter and larger than the upper. In introducing a probe the lid should be drawn outward, so as to obliterate the angle of the canal and convert it more nearly into a straight tube.

The *lachrymal sac* occupies a position at right angles to the lachrymal canals. It is crossed at its middle by the tendo oculi, and this tendon may be made apparent to the touch by drawing the lids outward. A knife entering below this tendon, and just within the edge of the orbit, would penetrate the sac. A probe introduced at this puncture, and passing downward, backward, and outward, would traverse the *nasal duct*, and appear in the inferior meatus of the nasal fossa.

The *orifice of the nasal duct* is to be found in the roof of the inferior meatus of the nasal fossa, beneath the inferior turbinated bone, about three quarters of an inch from the ala nasi, or, it is said, "at a distance equal to that between the inner angles of the eyelids." This duct may be explored from its nasal orifice by means of Gensoul's probe.

The *Eustachian tube* may be explored by a probe with a very short curve, carried along the inferior meatus of the nasal fossa, until it reaches beyond the hard palate; then turning the probe outward, it will enter the orifice, by which this tube opens into the pharynx behind the inferior turbinated bone. The finger passed into the mouth, and turned up behind the velum pendulum palati, will detect, upon the outer and upper wall of the pharynx, the cartilaginous lips, covered with mucous membrane, which characterize its opening.

The *parotid*, or *Steno's duct*, follows a course indicated by a line drawn from the tip of the lobe of the ear, forward, and nearly horizontally. This line crosses the masseter muscle, at the border of which the duct turns inward, to open in the mouth by an orifice in the cheek, opposite the first or second molar tooth. It passes in close proximity to the lower border of the malar bone. The transverse facial artery lies just above the duct, and an important branch of the facial nerve, which supplies the buccinator muscle, accompanies it along its upper edge.

The *antrum of the superior maxilla* may easily be reached by perforating the bone at the fossa on its anterior surface, just over the second bicuspid tooth, or, through the alveolar cavity from which a molar tooth has been extracted.

The *facial artery* crosses the lower jaw obliquely towards the angle of the lips, in front of the insertion of the masseter muscle, the outline of which is distinguishable in most subjects. In front of the angle of the inferior maxilla, the finger will detect a superficial notch, in which, with rare exceptions, the facial artery lies, and where its pulsations may be felt.

The *mental foramen*, from which issues the inferior dental branch of the fifth nerve, is situated a little nearer to the alveolar than the lower border of the inferior maxilla, at a point corresponding to the canine tooth, or to the interval between it and the bicuspid tooth. The supra and infra-orbital and mental foramina, are not always in a vertical line, one with the other, as is sometimes asserted.

To *explore the throat*, the tongue should be slightly protruded; the mouth being widely open, the posterior fauces and pharynx will be displayed by inspiring, and repeatedly pronouncing the syllable "hah."

The *glands of the tongue* may be advantageously studied upon the living subject.

The *ranine arteries* are situated at the bottom of the frenum linguæ, where it blends with the floor of the mouth. The large size of the *ranine veins*, lying upon the under surface of the tongue, should be noticed. On each side of the lower border of the frenum, the *Rivinian* and *Whartonian ducts* open. Their orifices in the centre of an obvious papilla may be plainly seen.

The *tonsil* is situated between the anterior and posterior pillars of the fauces. In its natural and healthy condition it hardly projects beyond these. The internal carotid artery lies at its base, and is separated from it by an intervening aponeurosis and by the constrictor muscle of the pharynx. This interval is, however, of considerable thickness. The angle of the inferior maxillary bone corresponds externally in its situation to that of the tonsil internally.

The *epiglottis* may be seen by depressing the tongue with a spatula.

The student should practice himself upon the dead subject in the *extraction of teeth*.

NECK.

The *external jugular vein* follows a course indicated by a line drawn from the angle of the lower jaw to the middle of the clavicle.

The *common carotid artery*, in the male, bifurcates at a point on a level with the upper border of the thyroid cartilage; in the female, opposite the middle of this cartilage.

The anterior border of the sterno-mastoid muscle covers the common, as well as the *external* and *internal carotid arteries*, the direction of which is indicated by a line drawn from midway between the anterior border of the mastoid process of the temporal bone and the ascending ramus of the lower jaw, to a point half an inch outside of the sterno-clavicular articulation. The common carotid is more deeply situated at the base of the neck than higher up, and its position in this part of its course corresponds to the interval between the sternal and clavicular attachments of the sterno-mastoid muscle. The artery may be compressed against the transverse processes of the cervical vertebræ. The posterior border of the sterno-mastoid muscle corresponds to the posterior border of the scalenus anticus muscle, which lies beneath it, and is, in part, the guide to the subclavian artery.

In the adult subject the *rings of the trachea* commence one and a half inches above the sternum; by throwing back the head an additional half inch may be exposed. The upper three rings are covered by the isthmus of the thyroid body. The trachea grows deeper as it descends, and at the base of the neck is sometimes an inch and a half from the surface. The *cricoid cartilage* can always be felt even in infants; it corresponds to the fifth cervical vertebra. The commencement of the œsophagus is immediately behind this cartilage.

The pulsations of the *subclavian artery* may be felt, deeply, behind the clavicle, in the interval between the posterior border of the sterno-mastoid and the anterior border of the trapezius muscles. By pressure downward and backward, the artery may be compressed against the first rib, which it crosses. The interval between the above-named muscles may be obliterated by their great degree of development.

The not unfrequent extension of the *pleural cavity* above the clavicle, sometimes to an extent of two and even three inches, will be made apparent on inflating the apex of the lung by a full and forced inspiration, and by percussion.

CHEST.

The *arteria innominata* corresponds in position to a line drawn from the centre of the union of the first with the second bone of the sternum, to the right sterno-clavicular articulation; by extending the neck its pulsations may be felt.

The *nipple*, in the male, lies upon the space between the fourth and fifth ribs.

The *coracoid process* of the scapula may be felt below the clavicle, in the interspace between the deltoid and pectoralis major muscles. The line of this interspace corresponds to that of the course of the axillary artery.

The *position of the heart* may be determined by percussion. " The *apex* pulsates between the fifth and sixth ribs, two inches below the nipple and one inch to its sternal side. The *aortic valves* lie behind the third intercostal space to the left of the sternum. The *pulmonary valves* lie behind the junction of the third rib, on the left side, with the sternum. The *tricuspid valves* lie behind the middle of the sternum, about the level of the fourth costal cartilage. The *mitral valves* lie behind the third intercostal space, about one inch to the left of the sternum."

The *bifurcation of the trachea* corresponds in position to the line of union between the first and second bones of the sternum.

The *lower margin of the lung*, anteriorly, corresponds to the most depending portion of the sixth rib; laterally, to the eighth rib; posteriorly, to the tenth rib. It is obvious, therefore, that the pleural cavity may be penetrated without the lung being wounded.

BACK.

The *scapula* covers the ribs from the seventh to the tenth, inclusive.

The *bifurcation of the trachea* corresponds in position to the spine of the third dorsal vertebra.

The *kidney* lies in front of the last two or three ribs, and sometimes its lower extremity does not reach below the twelfth rib. Its position is outside the erector spinæ muscle.

The *cauda equina* commences at the second lumbar vertebra.

The *descending colon* may be opened without wounding the peritoneum, in an operation called *colotomy*, by a transverse incision, upon the left side, two fingers' breadth above the crest of the ilium, outside the erector spinæ muscle.

The prominences of the *posterior superior spinous processes of the ilium* should be noticed. Pressure during protracted confinement to the back in bed, frequently causes the integument over them to slough, and gives rise to "bedsores."

The *gluteal artery* emerges from the greater sacro-ischiatic foramen at a point which corresponds to the middle of a line, drawn from the posterior superior spinous process of the ilium to the upper border of the trochanter major.

ABDOMEN.

The *xiphoid cartilage* of the sternum corresponds in position to the tenth dorsal vertebra. Its variations in shape and direction, in different subjects, should be observed.

The pulsations of the *aorta* may readily be felt, in an emaciated person, through the abdominal parietes. It may be compressed against the vertebral column. The *bifurcation of the aorta* corresponds to a point just below, and a little to the left of, the umbilicus, and to the third lumbar vertebra.

The median line of the abdomen, which is sometimes represented by a sulcus, corresponds to the *linea alba*. The *lineæ arcuatæ*, along the outer border of the recti muscles, may be made apparent by throwing these into a state of contraction. The trocar may be thrust through either of these aponeurotic intervals in the operation of paracentesis abdominis; in the first named, just below the umbilicus; in the second, at a point midway between the umbilicus and the anterior superior spinous process of the ilium.

The *epigastric artery* pursues a course, indicated by a line drawn from the middle of Poupart's ligament to a point just above the umbilicus.

The *external abdominal ring* may be explored, and penetrated to a variable degree, by the forefinger, invaginating the scrotum, and carried up beneath the abdominal integument, in a direction upward and outward, along the side of the spermatic cord, which serves as a guide to the position of the ring.

The *regions of the abdomen* are indicated by arbitrary lines, viz: two vertical lines, each, from the most dependent portion of the cartilages of the eighth ribs to the centre of Poupart's ligament; a transverse line, corresponding to the summits of the ilia; a second transverse line, corresponding to the cartilages of the ninth ribs. We thus have three zones, each subdivided into three regions. These are named, in the upper zone, the right and left hypochondriac, and in the centre, the epigastric; in the middle zone, the right and left lumbar, and in the centre, the umbilical; in the lower zone, the right and left inguinal, and in the centre, the hypogastric. The parts which, within the abdomen, correspond to these regions may be tabulated as follows:—

R. Hypochondriac.	Epigastric.	L. Hypochondriac.
Right lobe of liver, and gall-bladder; upper part of ascending colon; upper part of right kidney; right supra-renal capsule.	Middle and pyloric end of stomach; left lobe of liver; cœliac axis; semi-lunar ganglion; pancreas; aorta; vena cava inferior.	Cardiac end of stomach; spleen; head of pancreas; upper part of descending colon; upper part of left kidney; left supra-renal capsule.
R. Lumbar.	Umbilical.	L. Lumbar.
Ascending colon; lower part of right kidney; small intestines.	Transverse colon; duodenum; great omentum; mesentery; small intestines.	Descending colon; lower part of left kidney; small intestines.
R. Inguinal.	Hypogastric.	L. Inguinal.
Cæcum and appendix.	Small intestines; bladder, when distended; uterus, in female.	Sigmoid flexure of colon.

Directly above the pubes the bladder may be reached, especially when distended, and may be punctured, without fear of wounding the peritoneum, which, in being reflected from the abdominal parietes to the pelvic viscera, leaves quite an interval above the pubic bones.

The *vas deferens* may be felt as a constituent part of the spermatic cord; it is hard and round, usually situated near its posterior surface, and rolls between the finger and thumb searching for it in the substance of the cord.

The *epididymis* can be felt as a dense, oblong body, situated at the upper and back part of the testis; its position is, however, variable. It may lie in apposition with the lower border of the testis, and constitute what has been called "inversion of the testis." Its relations with the testis and with the spermatic cord should be carefully appreciated.

The external conformation of the *genital organs* should be studied; especially the variable degree of development characterizing the *frenum preputii*, and the greater dependence of the left testicle than the right, by which their mobility is increased and their liability to be compressed, one against the other, by the thighs, is diminished. The introduction of the catheter should be practised.

In the female the *vulva* is to be examined, and the vaginal exploration of the os uteri practised. The meatus of the female urethra is situated about half an inch below the clitoris, just above the protruding border of mucous membrane forming the orifice of the vagina. The lips of the meatus offer a small tubercle to the touch. The introduction of the female catheter, without the aid of sight, should be repeatedly performed.

The *prostate* may be felt, as a rounded and dense body, shaped like a chestnut, situated at the neck of the bladder, on introducing the fore finger, oiled, into the rectum, as far up as it can reach. It lies between the rectum and the symphysis pubes. Its dimensions vary, and measurements, as regards its size, are arbitrary. In early life it is small, in old age it is hypertrophied. Behind this organ, in the median line, the over-distended bladder may be punctured through the rectum.

UPPER EXTREMITY.

The *greater tuberosity of the humerus* is on a line with the external condyle of the same bone.

The *cephalic vein*, and the *inferior acromial* branch of the *thoracica acromialis* artery, lie in the depression which marks the interval between the pectoralis major and deltoid muscles.

The *axillary artery* traverses the axilla nearer to the anterior than the posterior border of that space. Dividing the axilla, longitudinally, into thirds, the line of union between the anterior and middle thirds indicates its exact position. Its pulsations may be felt, and the artery may be compressed against the head of the humerus.

The pulsations of the *brachial artery*, which commences at the lower border of the tendon of the latissimus dorsi muscle, may be felt at the inner edge of the coraco-brachialis and biceps muscles. At the bend of the elbow it occupies a position midway between the tips of the two condyles. The artery may be compressed against the shaft of the humerus.

The *musculo-spiral nerve* lies in the depression above the bend of the elbow, marking the interval between the outer border of the brachialis anticus and biceps muscles, and the inner border of the supinator longus muscle.

The *ulnar nerve* lies behind the inner condyle of the humerus, in the depression between it and the olecranon, being in close relation to the side of the latter process.

By rotating the arm at the wrist the *head of the radius* may be felt near the elbow, rolling under the finger which searches for its position. The forearm should be flexed. It then lies just in front of the external condyle.

The *veins at the bend of the elbow* may be demonstrated by the application of a ligature tied around the arm above, their consequent distension rendering them obvious to the eye and touch.

The *ulnar artery*, in the upper fourth of its course, is indicated by a line drawn from the middle of the elbow obliquely inward, and thence, by a line drawn from the tip of the internal condyle to the inner border of the pisiform bone.

The *radial artery* follows a course indicated by a line drawn from the middle of the bend of the elbow to the interval at the wrist between the tendons of the flexor communis digitorum and that of the supinator longus muscle. The pulsations of the *superficialis volæ* branch of this vessel, when it is present, may be felt upon the ball of the thumb, the muscles of which it crosses, near their origin, in a direction that continues the line of the radial artery.

The "*anatomist's snuff-box*" is the triangular interval between the tendon of the extensor secundi internodii pollicis, and the parallel tendons of the extensor ossis metacarpi pollicis and the extensor primi internodii pollicis muscles. Forced abduction of the thumb will reveal the depression to which the above name has been given. The radial artery traverses this space in the direction of a line drawn from the tip of the styloid process of the radius to the head of the metacarpal bone of the forefinger. Its pulsations may be felt.

The position of the *superficial palmar arch* nearly corresponds to a line drawn across the palm from the bottom of the cleft of the thumb, or to that crease in the palm which runs obliquely and transversely, and which is nearest to the carpus. The *deep palmar arch* is situated posteriorly to the superficial arch.

LOWER EXTREMITY.

The pulsations of the *external iliac artery* may be felt just above the middle of Poupart's ligament, and at this point the artery may be compressed against the horizontal ramus of the os pubis.

Poupart's ligament may be felt as a rounded cord, most distinct at its inner extremity, extending from the anterior superior spinous process of the ilium to the spine of the pubes; the spermatic cord crosses it obliquely at its inner end, which constitutes the inferior pillar of the external abdominal ring.

The general outlines of *Scarpa's triangle*, unless concealed by adipose tissue, may be seen in the upper and anterior part of the thigh. It is limited by Poupart's ligament above; by the upper border of the adductor longus muscle on its inside, and by the upper border of the sartorius muscle on the outside. Its apex is about four inches below Poupart's ligament. At this point the pulsations of the femoral artery may be felt, and it is here that the tourniquet is usually applied in amputations of the lower extremity.

The *saphenous opening* may be felt, in a thin subject, just below Poupart's ligament. Its outer border corresponds to the inner edge of the femoral artery. An enlarged gland sometimes conceals it.

The *femoral artery* follows a course indicated by a line drawn from the middle of Poupart's ligament to the posterior edge of the inner condyle. At the lower fourth of the thigh, upon the inside, a tendinous cord may be felt along its inner border. This is formed by the outer border of the tendon of the adductor magnus muscle, and con-

23*

tains the aperture called Hunter's canal, through which the femoral artery passes to reach the popliteal space. This orifice may sometimes be felt by the finger.

The *patella* rests upon the condyles of the femur. The tendon of the triceps extensor muscle, and the ligament of the patella below, divide the articulation of the knee into two lateral halves, and the expansion of the patella laterally, divides it, to a certain extent, transversely. If there is any increase of the synovial fluid, the fulness of the capsule will be most obvious above and below the patella, on each side of the tendon and of the ligament. The articulation will consequently have a quadrilateral shape, with four projections corresponding to the above-mentioned localities.

The *trochanter major*, when the thigh is rotated, describes an arc of a circle, the radius of which is equal to the length of the head and neck of the femur. The inner condyle being on the same plane as the head of the bone, indicates always the direction which the latter assumes.

The *sciatic nerve* follows a course indicated by a line drawn from a point, midway between the trochanter major and the tuberosity of the ischium, to the upper angle of the popliteal space.

The *popliteal space* is a diamond-shaped depression at the posterior aspect of the knee-joint. Its lower sides are formed by the prominences of the two separated muscular bellies by which the gastrocnemius muscle arises. Its upper and outer side is formed by the lower extremity of the biceps muscle. Its upper and inner side is formed by the following four muscles, enumerated in their order from within outward; semi-membranosus, semi-tendinosus, gracilis, and sartorius.

The *popliteal artery*, extending from the opening in the tendon of the adductor magnus, to the upper border of the soleus muscle, follows a line, a little oblique, from within outward, but corresponding, in a general way, to the long axis of the popliteal space. It is, however, a trifle nearer its inner than its outer side. The artery lies too deep for its pulsations to be felt. It should be noticed that when the leg is flexed, a bullet might traverse the popliteal space, or the ham-strings be divided, without wounding the popliteal artery; but that this could not happen when the limb is straight and extended as in the erect position.

The *anterior tibial artery*, in the leg, follows a course

indicated by a line drawn from the tubercle, to be felt on the side of the head of the tibia, to a point at the ankle, midway between the malleoli.

The pulsations of the *posterior tibial artery* may be felt at a point midway between the posterior border of the inner malleolus and the tendo Achillis.

The *tendon of the tibialis anticus* lies upon the instep, and is the innermost of all the tendons of the dorsum of the foot. It may be felt between the ankle and the inner border of the metatarsal bone of the great toe. This tendon is often divided, subcutaneously, for the relief of deformity.

The *tibialis posticus tendon* may be reached behind the inner malleolus, at a point midway between the anterior and posterior border of the foot. This tendon is also often divided, subcutaneously, for the relief of deformity.

The *dorsalis pedis artery* follows a course indicated by a line drawn from a point, midway between the malleoli, to the base of the first interosseous space.

The prominence of the head of the first metatarsal bone may be felt at the inner border of the foot. On its outer border the projecting tubercle of the head of the fifth metatarsal bone may also be felt and seen. These are important landmarks in the operation, called Lisfranc's, for disarticulating the foot at its tarso-metatarsal articulation. Posterior to the point at which the head of the first metatarsal bone is found, may be felt the projecting tubercle of the scaphoid bone; this is a landmark in disarticulating the foot through the middle of the tarsus, in what is known as Chopart's operation.

INDEX.

	PAGE
Abdomen,	162
arteries of,	176
superficial fascia of,	163, 168
Abdominal regions,	173, 266
Acervulus,	91
Acini,	190
Allantois,	213
Ampullæ,	98
Anatomist's snuff-box,	149
Andersch, ganglion of,	37, 52
Annulus albidus,	77
ovalis,	117
Anti-helix,	14
Anti-tragus,	14
Aorta, abdominal,	177
arch of,	56, 114
ascending,	114
descending,	114
sinuses of,	119
thoracic,	121
transverse,	114
Aponeurosis, epicranial,	16
pharyngeal,	61
temporal,	25
vertebral,	128
Apparatus ligamentosus colli,	75
Appendices epiploicæ,	184
Appendix of the auricles,	115
testicle,	218
vermiformis cæci,	183
Aqueduct of Sylvius,	91
Aqueductus cochleæ,	99
vestibuli,	97
Aqueous humor,	78
Arachnoid of the brain,	79
spinal cord,	131
Arbor vitæ of the cerebellum,	94
uterus,	221
Arch, aortic,	56, 114
crural,	197
palmar, deep,	155
superficial,	152
plantar,	248
Arches of the palate,	63
Arciform fibres,	84
Arnold's ganglion,	49, 64
Artery, accessory hepatic,	178
acromial, inferior,	104
anastomotica magna, of arm,	107
of thigh,	228
angular,	23
aorta, abdominal,	177
arch of,	56

	PAGE
Artery—aorta,	
thoracic	121
articular of the hip,	229
knee,	236
auditory, internal,	99
auricular, anterior,	14, 45
posterior,	14, 45
axillary,	103, 104
azygos articular,	237
basilar,	80
brachial,	106
bronchial,	121
buccal,	50
bulb, of the,	204, 205
calcanear,	243
carotid, common,	42, 44, 56
external,	44
internal,	30, 52, 81
temporal portion of,	65
carpal, radial,	143, 149
ulnar,	144
centralis retinæ,	32
cerebellar, inferior,	80
superior,	80
cerebral, anterior,	81
middle,	81
posterior,	80
anterior cervical,	58
ciliary,	31, 32
circumflex, anterior,	104
external,	227, 230, 234
superficial,	195
internal,	228, 234
ilii,	166, 201
posterior,	104
cœliac,	177
colica, dextra,	179
media,	179
sinistra,	179
comes nervi ischiatici,	231
phrenici,	111
communicating, anterior,	81
of the foot,	240
hand,	152
posterior,	81
coronary of the heart,	116, 119
lips,	23
cystic,	178
dental, inferior,	22, 36
superior,	50
digital of the foot,	240, 248
hand,	152
dorsalis, hallucis,	240

INDEX

Artery, dorsalis—
  nasi, 31
  pedis, 239, 240
  penis, 205, 209
  pollicis, 149
  scapulæ, 139
  epigastric, 167, 201
  superficial, 163, 195
    superior, 111
ethmoidal, 31
facial, 22, 44
femoral, 226
frontal, 18, 31
gastric, 178
gastro-duodenalis, 178
  epiploica dextra, 178
  sinistra, 178
gluteal, 209, 231
hemorrhoidal, external, 202
  middle, 208
  superior, 179, 209
hepatic, 178
  accessory, 178
hyoid, 44
hypogastric, 208, 256
ileo-colic, 179
iliac, common, 181
  external, 181, 201
  internal, 207
ilio-lumbar, 201, 209
infra-orbital, 19, 51
innominata, 56
intercostal, 121
  anterior, 111
  external, 121
  posterior branches of, 129
  superior, 59
interosseous, 144, 145
  anterior, 145
  of foot, 240
  of hand, 149, 155
  posterior, 145, 147
  recurrent, 148
ischiatic, 208, 231
labial, inferior, 22
lachrymal, 31
laryngeal, 44
lateral sacral, 209
lateralis nasi, 23
lingual, 44, 68
lumbar, 181
  posterior branches of, 129
malleolar, 240
mammary, external, 104
  internal, 59, 111
masseteric, 50
maxillary, internal, 50
median, 144
mediastinal, 111
meningea, anterior, 30
  inferior, 30, 45
  media, 30, 50
  parva, 30, 50
  posterior, 30, 58
mesenteric, inferior, 179
  superior, 178
metacarpal, 149
metatarsal, 240
musculo-phrenic, 111
mylo-hyoid, 50

Artery—
  nasal, 18, 31
  nutrient of femur, 228
    of tibia, 243
  obturator, 201, 228, 229
  occipital, 16, 45, 130, 131
  œsophageal, 121
  ophthalmic, 31
  orbital, 45, 51
  ovarian, 180, 209
  palatine, posterior, 50
  palpebral, 18, 31
  pancreatico-duodenalis, 178
  parotidean, 45
  perforating of the thigh, 228, 234
    hand, 155
    foot, 248
  pericardiac, 111
  perineal, superficial, 203
  peroneal, 243
    anterior, 240
  pharyngeal, ascending, 45, 53
  phrenic, 177, 194
  plantar, external, 246
    internal, 246
  popliteal, 236
  posterior scapular, 59, 129, 139
  princeps cervicis, 45, 130
    pollicis, 154
  profunda cervicis, 59, 129
    of the thigh, 227
    superior, 107
    inferior, 107
  pterygoid, 50
  pterygo-palatine, 50
  pudic external, superficial, 195
    deep, 195
    internal, 205, 208, 232
  pulmonary, 114
  pyloric, 178
  radial, 141, 142, 149, 154
  radialis indicis, 154
  ranine, 68
  recurrent interosseous, 148
    radial, 142
    tibial, 239
    ulnar anterior, 144
      posterior, 144
  renal, 180
  sacral, lateral, 209
  sacra media, 181
  scapular, posterior, 59, 129, 139
  sigmoid, 179
  spermatic, 180
  spheno-palatine, 50
  spinal, 121, 135
  splenic, 178
  stylo-mastoid, 45
  subclavian, 57
  submental, 45, 47
  subscapular, 104, 137, 138
  superficialis cervicis, 58, 129
    volæ, 143, 151
  supra-orbital, 31
  supra-renal, 180
  supra-scapular, 58, 129, 138
  sural, 236, 241
  tarsal, 240
  temporal, 25, 45
    anterior, 25, 45

INDEX.

Artery, temporal—
    deep, 50
    middle, 25, 45
    posterior, 25, 45
  thoracic, internal, 111
  thoracica acromialis, 104
    axillaris, 104
    inferior, 104, 137
    superior, 104
  thyroid inferior, 58, 43
    lowest, 58
    middle, 56, 58
    Neubauer, of, 58
    superior, 43, 44
  tibial anterior, 236, 239, 242
    posterior, 242, 246
  transversalis cervicis, 58
    perineī, 203
  transverse facial, 23, 45
  tympanic, 50
  ulnar, 144
  uterine, 209
  vaginal, 209
  vasa brevia, 178
  vertebral, 30, 58, 73, 80
    posterior branches of, 129
  vesical, inferior, 208
    middle, 208
    superior, 208
  Vidian, 50
Articulations, acromio-clavicular, 158
  alto-axoidean, 74
  carpo-metacarpal, 161
  costo-vertebral, 156
  coxo-femoral, 250
  metacarpo-phalangeal, 161
  occipito-atloid, 74
  phalangeal, 161
  sternal, 111
  sterno-clavicular, 54
  temporo-maxillary, 48
  tibio-femoral, 251
  tibio-tarsal, 253
Arytenoid cartilages, 71
Auditory canal, 94
Auricle, right, 116
  left, 118
Auriculo-ventricular openings, 117, 119
Axilla, 102
Axis cœliac, 177
  thyroid, 58

Bands, ventricular, 69
Bartholinus's glands, 220
Base of the brain, 85
Bichat, fissure of, 89
Bifurcation of the trachea, 115
Bladder, 207, 213
  fœtal, 257
Bone, hyoid, 70
Bone, spongy, 65
  turbinated, 65
Brain, 79, 81
Bronchi, 115
Brunner's glands, 185
Bulb of corpus spongiosum, 215
Bulbous portion of urethra, 217

Cæcum, 183
Calamus scriptorius, 93

Calyx of the kidney, 192
Canal, crural, 197
  Fontana, of, 77
  Hunter's, 227, 229
  inguinal, 169
  lachrymal, 17
  Nuck, of, 161, 258
  Petit, of, 78
  Schlemm, of, 77
  semicircular, 98
Capsule of Glisson, 190
  of the knee-joint, 251
Capsules, supra-renal, 191
Caput coli, 183
  gallinaginis, 216
Cartilage, alar, 21
  inter-articular of clavicle, 54
  jaw, 49
  pubes, 270
  wrist, 160
  lateral, 20
  of nose, 20
  of Santorini, 71
  semilunar, 253
  sesamoid, 21
  thyroid, 70
Caruncula lacrymalis, 18
Carunculæ myrtiformes, 220
Cauda equina, 135
Cava, vena, inferior, 111, 182
  superior, 55, 113
Cavity, pre-peritoneal, 167
  visceral, 172
Centrum ovale majus, 86
  minus, 86
Cerebellum, 92
Cerebrum, 85
Cerebro-spinal fluid, 80
Ceruminous glands, 15
Cervical ganglia, 53, 54, 59
Chambers of the eye, 78
Chiasma, 82
Chordæ tendineæ, 118
  vocales, 69
  Willisii, 27
Choroid, 76, 77
  plexus of fourth ventricle, 93
  lateral ventricle, 88
  third ventricle, 90
Ciliary ligament, 77
  processes, 77
Circle of Willis, 81
Circulation, fœtal, 258
Clitoris, 219
Cochlea, 98
Cœliac axis, 177
Colon, 183, 172
Columnæ carneæ, 118
Columns of Bertin, 193
  Morgagni, 213
  spinal cord, 136
  the vagina, 221
Commissure of spinal cord, 136
Commissures of brain, 90
Concha, 14
Conjoined tendon, 165, 166
Conjunctiva, 17
Conus arteriosus, 118
Convolution of corpus callosum, 86
Convolutions, 86

276   INDEX.

Cord, spermatic,	169	Ear, internal,	97
Cords, ventricular,	69	middle,	95
vocal,	69	Ejaculatory duct,	214
Cornea,	76	Eminentia collateralis,	88
Cornua of lateral ventricles,	87	teres,	93
Corona glandis,	216	Encephalon,	79, 81
Coronary valve,	117	Endocardium,	116
Corpora albicantia,	85	Epididymis,	218
Arantii,	118, 119	Epigastric region,	173
cavernosa,	215	Epiglottidean gland,	71
quadrigemina,	91	Epiglottis,	71
Corpus callosum,	85, 86	Eustachian tube,	62, 96
fimbriatum,	89, 222	valve,	117, 258
geniculatum,	91	Eye,	76
olivare,	84	Eyelids,	17
pyramidale,	84		
restiforme,	84	Falciform border,	196
rhomboideum,	94	Fallopian tubes,	222
spongiosum,	215	Falx cerebelli,	29
striatum,	87	cerebri,	27
Corpuscles of the spleen,	186	Fascia cervical,	38
Cortical substance of brain,	86	cribriform,	196
of kidney,	193	deep,	196
Covered band of Reil,	87	iliaca,	196
Cowper's glands,	201, 217	infundibuliform of inguinal canal,	169
Cranial nerves,	33, 82	of crural canal,	197
Cribriform fascia,	196	lata,	196
Cricoid cartilage,	71	leg of the,	238
Crico-thyroid membrane,	71	lumborum,	127
Crista of the vestibule,	97	obturator,	210
Crura cerebelli,	84, 92	palmar,	151
cerebri,	84	plantar,	215
of the diaphragm,	194	pelvic,	210
fornix,	89	perineal,	202
Crus penis,	205	propria,	198
Crural arch,	197	recto-vesical,	210
canal,	197	spermatic,	164, 169
septa of,	197	superficial of abdomen,	163, 168
ring,	197	perineum,	202
Crystalline lens,	78	thigh,	195
Cuneiform cartilages,	71	transversalis,	169
Cupola,	98	Femoral hernia,	196
Cystic duct,	176, 190	Fenestra ovalis,	95
		rotunda,	95
Dartos,	217	Fibres, arciform,	84
Detrusor urinæ,	213	inter-columnar,	164, 169
Decussation of anterior pyramids,	84	Fibro-cartilage, inter-articular of the	
Diaphragm,	193	clavicle,	54
larger muscle of the,	194	of the jaw,	49
lesser,	194	knee,	233
Duct, cystic,	176, 190	nose,	20, 21
hepatic,	176, 190	pubes,	230
lactiferous,	101	wrist,	160
nasal,	18, 66	Filum terminale,	134
pancreatic,	187	Fimbriated extremity of Fallopian	
prostatic,	216	tube,	222
Rivinian,	66	Fissura Glaseri,	95
Steno's,	24	Fissure of Bichat,	89
thoracic,	55, 123, 195	of Sylvius,	81, 85
Wharton's,	47	transverse of the brain,	89
Ductus arteriosus,	114, 258	Fissures of the spinal cord,	136
choledochus communis,	176, 190	liver,	188
communis ejaculatorius,	214	Floating kidney,	193
lymphaticus dexter,	55, 123	Fluid, sub-arachnoid,	80, 134
venosus,	257	Fœtal circulation,	258
Duodenum,	183, 185	Fœtus, anatomy of,	256
Dura mater of the brain,	26	Fold, ary-epiglottidean,	69
spinal cord,	134	recto-vesical,	205
		recto-uterine,	206
Ear, external,	13, 41	semilunar of Douglass,	167

INDEX. 277

Fold—	
vesico-uterine,	206
Follicles of Lieberkuhn,	184
Foramen cæcum, of brain,	84
tongue,	67
commune anterius,	90
posterius,	90
Munro, of,	90
ovale,	258
Soemmering, of,	78
Winslow, of,	174
Foramina Thebesii,	117
Fornix,	89
Fossa, inguinal,	170
innominata,	14
ischio-rectal,	205
nasal,	65
navicularis,	217
ovalis,	117
scaphoid,	14
Fourchette,	219
Fovea hemispherica,	97
semi-elliptica,	97
Frenum epiglottidis,	66
of lips,	60
tongue,	66
prepuce,	216
Gall-bladder,	190
Ganglia, cervical,	53, 54, 59
lumbar,	181
sacral,	207
thoracic,	122
Ganglion, Andersch, of,	37, 52
Arnold's,	49, 64
Gasserian,	35
cervical, inferior,	54, 59
middle,	53
superior,	53
jugulare,	36
lenticular,	34
Meckel's,	64
otic,	49, 64
petrosum,	37, 52
of the root,	37
semilunar,	177
spheno-palatine,	64
submaxillary,	47
thyroid,	53
Genital organs,	267
Genu of the corpus callosum,	85
Gimbernat's ligament,	196
Gland, agminated,	184
Bartholinus's,	220
bronchial,	115
Brunner's,	185
ceruminous,	15
Cowper's,	204
duodenal,	185
epiglottidean,	71
inguinal,	163, 196
labial,	21
lachrymal,	18
Lieberkuhn's,	184
mammary,	101
meibomian,	17
mesenteric,	179
Pacchionian,	26
parotid,	24
Peyer's,	184

Gland—	
pineal,	91
pituitary,	29, 55
salivary,	46, 66
socia parotidis,	24
solitary,	184
sublingual,	66
submaxillary,	46
synovial of Havers,	251
thymus,	112, 258
tracheal,	71
Tyson's,	216
vulvar,	220
Glans, clitoridis,	219
penis,	216
Glisson's capsule,	190
Graafian vesicles,	222
Gubernaculum testis,	257
Guthrie's muscle,	210
Gyri,	86
Gyrus fornicatus,	86
Hamulus laminæ spiralis,	98
Heart,	113, 115
fœtal,	258
Helicotrema,	98
Holix,	14
Hemispheres of the brain,	86
Hepatic duct,	176, 190
Hey's ligament,	196
Hernia, congenital,	171
crural,	196
direct,	170, 171
encysted,	171
femoral,	196
inguinal,	168
oblique,	170, 171
cæcum, of the,	171
Hasselbach, triangle of,	171
Hilus of the kidney,	192
liver,	189
spleen,	186
Hippocampus major,	88
minor,	88
Horner's muscle,	18
Horseshoe kidney,	193
Hunter's canal,	227, 229
Hyaloid membrane,	79
Hyoid bone,	70
Hymen,	220
Hypochondriac region,	173
Hypogastric region,	173
Ileo-cæcal valve,	183
Ileum,	183
Incisura cerebelli,	92
Incus,	96
Infundibulum of the brain,	85
heart,	118
kidney,	192
Inguinal canal,	169
fossæ,	170
region,	163, 173
Inter-articular cartilage of the clavicle,	54
jaw,	49
wrist,	160
Inter-columnar fibres,	164, 169
Inter-peduncular space,	85
Inter-vertebral substance,	157

24

Intestinal tube,	183
Intestines, fœtal,	257
Intumescentia gangliformis,	36
Iris,	77
Ischio-rectal fossa,	202
Island of Reil,	85, 86
Isthmus faucium,	62
Iter ad infundibulum,	90
a tertio ad quartum ventriculum,	91, 93
Jejunum,	183
Joint, ankle,	253
elbow,	159
hip,	250
lower jaw,	48
knee,	251
shoulder,	159
wrist,	160
Kidneys,	191
fœtal,	257
pelvis of,	192
Labia majora,	219
minora,	220
Labyrinth,	97
membranous,	99
Lachrymal canal,	17
gland,	18
punctum,	18
sac,	18
Lacteals,	185
Lacuna magna,	217
Lamina cinerea,	85
cribrosa,	76
spiralis,	98
Landmarks,	260
Large intestine,	183
Larynx,	68
nerves of,	70
arteries of,	70
ventricles of,	69
Lateral tract,	84
ventricles,	87
Lens, crystalline,	78
suspensory ligament of,	78
Lenticular ganglion,	34
Lieberkuhn, follicles of,	184
Ligamenta brevia of the fingers,	154
toes,	247
subflava,	158
Ligaments, acromio-clavicular,	158
alar,	252
ankle, of the,	253
annular, anterior, of ankle,	238
external,	244
internal,	244
wrist, anterior of the,	147, 153
posterior of the,	149
anterior of knee,	252
of ankle,	253
of elbow,	159
of wrist,	160
arcuatum externum,	194
internum,	194
astragalo-scaphoid,	254
atlo-axoid,	74
bladder, of the,	207
broad, of the liver,	172
Ligaments—	
broad, of the uterus,	207, 221
calcaneo-astragaloid	254
cuboid,	254
scaphoid,	254
capsular of the articular processes,	158
hip,	250
knee,	251
shoulder,	159
thumb,	161
carpal,	160
carpo-metacarpal,	161
check,	75
chondro-sternal,	111
common, anterior,	157
posterior,	157
coccyx, of the,	249
conoid,	158
coraco-acromial,	158
clavicular,	158
humeral,	159
coronary, of liver,	188
costal,	111
costo-clavicular,	54
transverse,	156
vertebral,	156
xiphoid,	112
cotyloid,	251
crico-thyroid,	71
crucial,	75
of knee,	252
deltoid,	253
elbow, of the,	159
fibula, of the,	253
Gimbernat's,	196
glosso-epiglottidean,	66, 71
glenoid,	159
Hey's	196
hip-joint, of the,	250
hyo-epiglottic,	71
ilio-femoral,	250
lumbar,	249
interarticular of hip-joint,	251
of ribs,	156
interclavicular,	54
interosseous, fibula, of the,	253
metacarpal bones, of the,	161
metatarsal bones, of the,	254
middle,	157
transverse,	157
interspinous,	158
intertransverse,	158
jaw, of the,	48
knee, of the,	251
larynx, of the,	71
lateral, of the ankle,	253, 254
elbow,	159
jaw,	48, 49
knee,	252
phalanges, foot,	254
hand,	161
wrist,	160
liver, of the,	188
lumbo-sacral,	249
metacarpo-phalangeal,	161
metatarsal,	254
mucous,	252
nuchæ,	126
oblique,	160

INDEX. 279

Ligaments—
  obturator, 250
  occipito-atloid, 74
  axoid, 75
  odontoid, 75
  orbicular, 159
  ovary, of the, 222
  palpebral, 17
  patellæ, 252
  pelvis, of the, 249
  peroneo-tibial, 252
  phalangeal, of the foot, 255
    hand, 161
  plantar, 254
  posterior, of elbow, 159
    of wrist, 160
  posticum Winslowii, 235, 252
  Poupart's, 164
  pterygo-maxillary, 61
  pubic, 250
  Retzius, of, 238
  round, of the liver, 172
    uterus, 221
  sacro-coccygean, 249
  sacro-iliac, 249
  sacro-ischiatic, greater, 249
    lesser, 250
  shoulder, of the, 159
  stellate, 156
  sternal, 111
  sterno-clavicular, 54
  stylo-hyoid, 51, 61
  stylo-maxillary, 49
  sub-pubic, 250
  supra-condyloid, 159
  supra-spinous, 158
  tarsal, 254
  tarso-metatarsal, 254
  teres, 251
  thyro epiglottic, 71
  thyro-hyoid, 70
  transverse of the crucial, 75
    acetabulum, 251
    atlas, 75
    fingers, 151
    knee, 253
    metatarsus, 255
    scapula, 159
    semilunar cartilages, 253
  trapezoid, 158
  triangular, 168, 204
  tympanum, of the, 97
  uterus, broad, of the 207, 221
  wrist, of the 160
Ligamentum, denticulatum, 135
  breve plantæ, 254
  longum plantæ, 254
  suspensorium, of odontoid process, 75
    of penis, 163
  nuchæ, 126
Limbus luteus, 78
Linea alba, 163
  arcuata, 164
Lineæ transversæ, of fourth ventricle, 93
  rectus muscle, 166
Liquor Cotunnii, 99
  Scarpæ, 99
Lithotomy, 205
Liver, 187
  fœtal, 257

Lobes of the brain, 81
  liver, 188
  prostate, 214
  lungs, 124
Lobule of the ear, 14
Lobus caudatus, 189
  quadratus, 189
  Spigelii, 189
Locus niger, 94
  perforatus, 85
Lumbar fascia, 127
  region, 173
Lungs, 124
  fœtal, 258
Lymphatic glands of abdomen, 179
  axilla, 102
  bronchial, 115
  of elbow, 110
  inguinal, 163, 196
Lyra of the fornix, 89

Malleus, 96
Mammary gland, 101
Meatus auditorius, 14, 15
  of the urethra, female, 220
  male, 216
Meatuses of the nares, 65
Meckel's ganglion, 64
Meconium, 257
Mediastinum, anterior, 112
  posterior, 120
  testis, 218
Medullary substance of brain, 86
  of cerebellum, 93
Medulla oblongata, 83
Meibomian glands, 17
Membrana dentata, 135
  nictitans, 18
  tympani, 94
  secondary, 95
Membrane, choroid, 77
  costo-coracoid, 102
  crico-thyroid, 71
  of Descemet, 76
  Demours, 76
  hyaloid, 79
  interosseous, of forearm, 160
  of leg, 253
  obturator, 250
  pituitary, 66
  Schneiderian, 66
  thyro-hyoid, 68
Membranous urethra, 217
  labyrinth, 97
Mesenteric glands, 179
Mesentery, 174
Meso-colon, transverse, 174
Meso-cæcum, 174
Mesorchium, 258
Meso-rectum, 174
Middle ear, 95
Mitral valve, 119
Modiolus, 96
Mons Veneris, 219
Morsus diaboli, 222
Muscles,
  abductor indicis, 150
    minimi digiti pedis, 246
    manus, 152
    pollicis pedis, 246

INDEX.

Muscles—
- abductor pollicis manus, 151
- acceleratores urinæ, 203
- accessorius, 247
-     ad sacro-lumbalem, 130
- adductor, brevis, 228
-     longus, 226
-     magnus, 228
-     minimi digiti manus, 152
-     pollicis pedis, 247
-         manus, 154
- anconeus, 147
- anti-tragicus, 15
- arytenoideus, 68
- attollens aurem, 14
- attrahens aurem, 13
- azygos uvulæ, 63
- basio-glossus, 67
- biceps, of the thigh, 234
-     arm, 105, 142, 160
- biventer cervicis, 131
- brachialis anticus, 106, 142
- buccinator, 22
- bursalis, 233
- cerato-glossus, 67
- cervicalis ascendens, 131
- chondro-glossus, 67
- ciliary, 77
- coccygeus, 212
- complexus, 131
- compressor nasi, 20
-     urethræ, 211
- constrictor isthmi faucium, 64
-     medius, 61
-     inferior, 61
-     superior, 61
-     vaginæ, 205
- coraco-brachialis, 106
- corrugator supercilii, 17
- cremaster, 165
- crico-arytenoideus lateralis, 69
-     posticus, 68
-     thyroideus, 68
- cruræus, 225
- deltoid, 101
- depressor anguli oris, 21
-     labii inferioris, 21
-     supercilii, 17
-     labii superioris alæque nasi, 20
- detrusor urinæ, 213
- diaphragm, 193
- digastricus, 46
- ear, of the, 13
- erector clitoridis, 205
-     penis, 203
-     spinæ, 130
- extensor carpi radialis brevior, 146
-     longior, 146
-     ulnaris, 147
-     minimi digiti, 14·
-     brevis digitorum, 239
-     communis digitorum, 147
-     longus digitorum, 238
-     indicis, 140
-     ossis metacarpi pollicis, 148
-     proprius pollicis, 238
-     primi internodii pollicis, 148
-     secundi internodii pollicis, 148
- flexor brevis digiti minimi, 152
-     pedis, 248

Muscles, flexor—
- brevis pollicis, 151
-     pedis, 247
- carpi radialis, 141
-     ulnaris, 141
- brevis digitorum, 245
-     profundus, 145, 254
-     sublimis, 142, 153
- longus digitorum, 243, 247
-     pollicis, 145, 151, 154
-     pedis, 243, 247
- ossis metacarpi pollicis, 151
-     minimi digiti, 152
- gastrocnemius, 241
- gemellus, inferior, 233
-     superior, 233
- genio-hyo-glossus, 67
- genio-hyoid, 47
- glosso-pharyngeus, 61
- gluteus maximus, 230
-     medius, 230
-     minimus, 232
- gracilis, 226
- Guthrie's, 210
- helicis, major, 15
-     minor, 15
- Horner's, 18
- hyo-glossus, 67
- iliacus internus, 200, 226
- infra-costales, 124
- infra-spinatus, 138
- intercostal, external, 124
-     internal, 124
- interosseous, dorsal, of the hand, 150
-     foot, 240
-     palmar, 155
-     plantar, 248
- interspinales, 133
- intertransversales, 73, 133
- larynx, of the, 68
- latissimus dorsi, 126
- laxator tympani, 97
- levator anguli oris, 20
-     scapulæ, 129
-     ani, 211
-     glandulæ thyroideæ, 43
-     labii superioris, 19
-         alæque nasi, 19
-         inferioris, 22
- levator palati, 63
-     palpebræ, 32
- levatores costarum, 132
- lingualis, inferior, 67
-     superior, 67
- longissimus dorsi, 131
- longus colli, 72
- lumbricales of the foot, 247
-     hand, 154
- marsupialis, 233
- massa carnea Jacobi Sylvii, 247
- masseter, 25
-     internal, 48
- multifidus spinæ, 132
- mylo-hyoid, 47
- obliquus externus, 163
-     internus, 165
-     inferior, 32, 133
-     superior, 32, 133
- obturator externus, 229, 233
-     internus, 212, 232

INDEX.

Muscles—
- occipito-frontalis, 15, 131
- omo-hyoid, 40, 129
- opponens minimi digiti, 152
- pollicis, 151
- orbicularis oris, 21
- palpebræ, 16
- palato-glossus, 64
- pharyngeus, 63
- palmaris brevis, 150
- longus, 141
- pectineus, 226
- pectoralis major, 100
- minor, 102
- peroneus brevis, 244
- longus, 214, 248
- tertius, 239
- plantaris, 242
- platysma myoides, 38
- popliteus, 243
- pronator quadratus, 145
- radii teres, 141
- psoas magnus, 200, 226
- parvus, 200
- pterygoid, external, 48
- internal, 48
- pyramidalis, of the abdomen, 167
- nasi, 16
- pyriformis, 212, 232
- quadratus femoris, 233
- lumborum, 200
- menti, 21
- quadriceps extensor cruris, 225
- rectus abdominis, 166
- capitis anticus major, 72
- minor, 72
- lateralis, 72
- posticus major, 133
- minor, 133
- femoris, 224
- external, of the eye, 32
- inferior, 32
- internal, 32
- superior, 32
- sternalis, 167
- retrahens aurem, 14
- rhomboideus major, 128
- minor, 127
- risorius Santorini, 22
- sacro-lumbalis, 130
- salpingo-pharyngeus, 62
- sartorius, 224
- scalenus anticus, 73
- medius, 73
- posticus, 73
- semi-spinalis, 132
- colli, 132
- dorsi, 132
- semi-membranosus, 234
- semi-tendinosus, 234
- serratus magnus, 137
- posticus inferior, 128
- superior, 128
- soleus, 242
- sphincter ani, 203
- internus, 203
- vesicæ, 213
- spinalis dorsi, 131
- splenius, 128
- capitis, 128

Muscles—
- splenius colli, 128
- stapedius, 97
- sterno-hyoid, 42
- sterno-mastoid, 40
- sterno-thyroid, 43
- stylo-glossus, 51
- stylo-hyoid, 46
- stylo-pharyngeus, 51, 61
- subclavius, 102
- sub-cruræus, 225
- subscapularis, 137
- supinator brevis, 148
- longus, 142
- supra-spinatus, 138
- temporal, 26
- tensor palati, 63
- tarsi, 18
- tympani, 97
- vaginæ femoris, 224
- teres major, 138
- minor, 139
- thyro-arytenoideus, 69
- thyro-hyoid, 43
- tibialis anticus, 238
- posticus, 244, 248
- trachelo-mastoid, 131
- tragicus, 15
- transversalis of the abdomen, 165
- colli, 131
- pedis, 248
- transverse, of the ear, 15
- transversus perinei, 203, 205
- alter, 203
- trapezius, 126
- triangularis menti, 21
- sterni, 110
- triceps extensor cruris, 225
- cubiti, 139
- trochlearis, 32
- ureters of the, 213
- vastus externus, 224
- internus, 225
- Wilson's, 211
- zygomaticus major, 20
- minor, 20
- Musculi pectinati, 117

Nares, 65
Nasal duct, 18, 66
fossæ, 65
Nates of the brain, 91
Nerves, abdominal, 177, 199
- abducens, 36, 82
- accessory obturator, 199
- spinal, 37, 41, 52, 83, 135
- acromial, 40
- auditory, 36, 83, 99
- auricularis magnus, 14, 39
- posterior, 14, 23
- auriculo-temporal, 24, 49
- brachial, 59, 102
- buccal, 24
- cardiac, 52, 120
- inferior, 59
- middle, 53
- superior, 53
- cervical anterior, 39
- posterior, 130
- cervico-facial, 23

282   INDEX.

Nerves—
  chorda tympani,   47, 50, 95
  circumflex,   108
  clavicular,   40
  coccygeal,   210
  cochlear,   36, 99
  communicans noni,   40, 42
    peronei,   236
    poplitei,   236
  cranial,   33, 82
  crural,   199, 223
  cutaneous, of the back,   130
    external, of arm,   107
      thigh,   199, 223, 234
    internal,   108
    lesser,   108
    middle posterior,   230, 234
    musculo,   107
    plantar,   243
  dental, inferior,   21, 50
    superior,   35
  descendens noni,   41, 47
  digastric,   46
  digital, of the foot,   237, 247
    hand,   146, 150, 153
  dorsal,   130
  dorsalis penis,   203, 215
  eighth pair,   36, 83
  facial,   23, 36, 83
  fifth pair,   35, 82
  first pair,   33, 82
  fourth pair,   34, 82
  frontal,   35
  gastric,   120
  genito-crural,   199, 223
  glosso-pharyngeal,   36, 52, 68, 83
  gluteal,   210, 231
  gustatory,   47, 50, 68
  hypo-glossal,   37, 47, 50, 68, 83
  inferior maxillary,   36
  infra-maxillary,   24
  infra-orbital,   19, 35
  intercostal,   122
  intercosto-humeral,   108, 123
  interosseous, anterior,   143, 145
    posterior,   108, 148
  Jacobson's,   37
  lachrymal,   35
  larynx, of,   70
  laryngeal, recurrent,   56, 70, 120
    superior,   52, 70
  lumbar,   130
  lumbo-sacral,   200, 210
  malar,   23
  masseteric,   49
  maxillary, inferior,   36
    superior,   35
  median,   108, 143, 151, 152, 153
  molles,   53
  motor oculi,   34, 82
  muscular, superior, of the bra-
                chial plexus,   102
    inferior,   102
  musculo cutaneous, of arm,   107
    leg,   237, 240
    thigh,   199, 166
  musculo-spiral,   108
  mylo-hyoid,   47, 50
  nasal,   35
  naso-palatine,   64

Nerves—
  ninth pair,   37, 83
  obturator,   199, 210, 228, 234
  occipitalis major,   14, 16, 130, 131
    minor,   14, 39
  œsophageal,   120
  olfactory,   33, 82
  ophthalmic,   35
  optic,   33, 82
  orbitar,   35
  palatine,   64
  palmar, superficial,   143
  par vagum,   37, 83
  perforans Casserii,   107
  perineal cutaneous,   203
  peroneal,   235
    cutaneous,   237
  petrosus superficialis major,   65
    minor,   64
  pharyngeal,   52
  phrenic,   40, 56, 112
  plantar, cutaneous,   243
    external,   246
    internal,   246
  pneumogastric, 37, 52, 55, 83, 120, 185
  popliteal,   236
  portio dura,   36
    mollis,   36
  pterygoid, external,   50
    internal,   50
  pudic, internal,   203, 210
  pulmonary,   120
  radial,   143, 146
  recurrent, laryngeal,   56, 70, 120
  renal,   103
  sacral,   130
  saphena, external,   236, 238, 241
    long,   223
    short,   223, 236
  sciatic,   210, 231, 235
    lesser,   210, 230, 234
  scrotal,   199
  second pair,   33, 82
  seventh pair,   36, 83
  sixth pair,   36, 82
  spermatic,   219
  spheno-palatine,   50
  spinal,   135
  spinal accessory,   37, 41, 52, 83, 135
  splanchnic, greater,   122
    lesser,   122
    renal,   122
  stylo-hyoid,   46
  sub-occipital,   73, 133
  subscapular,   103, 137
  superficialis colli,   39
  superior maxillary,   35
  supra maxillary,   24
    orbital,   17, 35
    scapular,   103, 129, 138
  sympathetic,   53, 122
    prevertebral portion of,   122
    vertebral portion of,   122
  temporal,   23
  temporo-facial,   23
  third pair,   34, 82
  thoracic, long,   103, 137
    short,   103
  tibial, anterior,   237, 240
    posterior,   237, 243

INDEX.

Nerves—
- trifacial, 35, 82
- trochlearis, 34, 82
- tympanic, 37
- ulnar, 108, 144, 146, 150, 155
- vestibular, 36, 99
- Vidian, 64
- Wrisberg, 108
- intermediate portion of, 83

Nose, cartilages of, 20
Nuck, canal of, 171, 258
Nymphæ, 220

Œsophagus, 62, 121
Omentum, gastro-splenic, 174
- great, 172, 174
- lesser, 174
Openings in the diaphragm, 194
Optic commissure, 82
- thalamus, 88, 91
- tract, 82
Oraserrata, 78
Orbicular process, 96
Orbiculare, os, 96
Orbit, arteries of, 31
- muscles, of, 32
- nerves of, 35, 36
Os orbiculare, 96
Os uteri, 221
- internum, 221
Otoconites, 99
Ovaries, 222
- fœtal, 258

Pacchionian glands, 26
Palate, pillars of, 63
- arches of, 63
- hard, 63
- soft, 63
Palmar arch, deep, 155
- superficial, 152
Palpebral ligaments, 17
Pancreas, 187
- lesser, 187
Papillæ of the tongue, 66
- caliciform, 67
- circumvallatæ, 67
- conical, 66
- filiform, 66
- fungiform, 66
- kidney, 192
Parietes of abdomen, 162
Parotid gland, 24
Par vagum, 37, 83
Peduncles of the cerebellum, 92
- corpus callosum, 85
- pineal gland, 91
Pelvis, viscera of, 206, 212
Penis, 215
Pericardium, 113
Perineal centre, 203
Perineum, 202
Peritoneum, 173, 205
- reflections traced, 174, 205
Pes anserinus, of the face, 24
- knee, 235
- hippocampi, 88
Peyer's glands, 184
Pharynx, 60
Pia mater of the brain, 80

Pia mater of the spinal cord, 134
Pillars of the palate, 63
- abdominal ring, 164, 160
- diaphragm, 194
Pineal body, 91
Pinna, 14
Pituitary body, 29, 85
- membrane, 66
Platysma myoides, 38
Pleura, 120, 264
Plexus, abdominal, 177
- aortic, 179
- brachial, 59, 102
- cardiac, 122
- deep, 116
- superficial, 116
- carotid, 31
- cavernous, 31
- cervical, anterior, 39
- posterior, 130
- choroid, 88
- hepatic, 190
- hemorrhoidal, 207
- hypogastric, 207
- lumbar, 199
- mesenteric, 179
- ovarian, 207
- phrenic, 177
- pneumogastric, 120
- prostatic, 207, 214
- pterygoid, 51
- pulmonary, 122
- renal, 180
- sacral, 210
- solar, 177
- spermatic, 180, 219
- splenic, 186
- supra-renal, 177, 191
- uterine, 207
- vaginal, 207
- vesical, 207
Plica semilunaris, 18
Pomum Adami, 70
Pons Tarini, 85
- Varolii, 84
Popliteal space, 235
Portal vein, 176
Portio dura, 36
- mollis, 36
Porus opticus, 78
Position of antrum, 262
- aorta, 265
- anatomist's snuff-box, 269
- arch, deep palmar, 269
- superficial palmar, 269
- artery, axillary, 268
- brachial, 268
- carotid, 263
- dorsalis pedis, 271
- epigastric, 266
- facial, 262
- femoral, 269
- gluteal, 265
- iliac, external, 260
- innominate, 264
- meningeal, middle, 260
- popliteal, 270
- radial, 268
- ranine, 262
- subclavian, 263

INDEX.

Position of—
- artery, superficialis volæ, 268
  - thoracic, inferior, 267
  - tibial, anterior, 271
  - posterior, 271
  - ulnar, 268
- articulation, medio-tarsal, 271
  - tarso-metatarsal, 271
- canal, lachrymal, 261
- cartilage, cricoid, 263
  - xiphoid, 265
- cauda equina, 265
- colon, 265
- duct, nasal, 261
  - parotid, 261
  - Rivinian, 262
  - Steno's, 261
  - Whartonian, 262
- epididymis, 267
- epiglottis, 267
- Eustachian tube, 261
- foramen, infra-orbital, 260
  - mental, 262
  - supra-orbital, 260
- heart, 264
  - valves of, 264
- ligament, Poupart's, 269
- linea alba, 265
  - arcuata, 265
- lung, margin of, 264
- nerve, musculo-spiral, 268
  - sciatic, 268
  - ulnar, 270
- nipple, 264
- opening, saphenous, 269
- patella, 270
- popliteal space, 270
- Poupart's ligament, 269
- process, coracoid, 264
  - posterior superior spinous of ilium, 265
  - prostate, 267
  - radius, head of, 268
  - ring, external abdominal, 266
  - sac, lachrymal, 261
- saphenous opening, 269
- scapula, 265
- Scarpa's triangle, 269
- sinus, lateral, 260
  - longitudinal, 260
- space, popliteal, 270
- tendon of tibialis anticus, 271
  - posticus, 271
- tonsil, 262
- trachea, bifurcation of, 264, 265
  - rings of, 263
- triangle, Scarpa's, 269
- trochanter major, 270
- tube, Eustachian, 261
- tuberosity of humerus, 267
- valves, cardiac, 264
- vein, cephalic, 267
  - external jugular, 263
  - ranine, 262
- vas deferens, 267
Poupart's ligament, 164
Preperitoneal cavity, 167
Prepuce of the clitoris, 219
  - penis, 215
Process, orbicular, 96

Process—
- supra-condyloid, 106, 141
- inferior vermiform, 92
- superior vermiform, 92
Processus auditorius, 91
  - cochleariformis, 96
  - ad medullam, 93
  - ad pontem, 92
  - e cerebello ad testes, 92
  - vaginalis peritonei, 238
Promontory, 95
Prostate, 214
Prostatic urethra, 216
Pulmonary artery, 114
  - plexus, 122
  - vein, 114
Punctum lacrymale, 18
Pylorus, 186
Pyramid, 90
Pyramids, anterior, 84
  - Malpighi, of, 192
  - posterior, 84

Receptaculum chyli, 194
Rectum 173, 212
Regions, abdominal, 173, 266
  - inguinal, 168
Reil, island of, 85
Rete pampiniforme, 219
Retina 78
Rima glottidis, 69
Ring, abdominal, external, 164
  - internal, 169
  - crural, 197
Rivinian ducts, 66
Root of the lung, 115, 125
Roots of the spinal nerves, 135
Rostrum, 85

Sac, lachrymal, 18
Saccule, 99
Sacculus laryngis, 69
Saphenous opening, 196
Scala tympani, 99
  - vestibuli, 99
Scarpa, triangle of, 226
Schneiderian membrane, 66
Sclerotic, 76
Scrotum, 217
Semicircular canals, 98
Semilunar fold of Douglass, 167
  - valves, 118, 119
Septa of crural canal, 197
Septum of the auricles, 118
  - ventricles, 119
  - cruralе, 197
  - lucidum, 89
  - nasi, 21, 65
  - of the tongue, 67
Sheath of the carotid artery, 41
  - flexor tendons, 153, 246
  - penis, 216
  - rectus muscle, 167
Sigmoid flexure, 173
  - valves, 118, 119
Sinuses of the aorta, 119
  - dura mater, 27
  - pulmonary artery, 118
  - uterus, 221
  - Valsalva, of, 119

INDEX.

285

Sinus, basilar,	29
cavernous	29
circular,	29
coronary,	116
lateral,	128
longitudinal, inferior,	29
superior,	29
occipital,	29
petrosal, inferior,	28
superior,	28
pocularis,	216
prostatic,	216
straight,	27
transverse,	29
Small intestines,	183
Soft palate,	63
Soemmering, foramen of,	78
Space, interpeduncular,	85
popliteal,	235
posterior perforated,	85
anterior perforated,	86
sub-arachnoid,	80, 134
Speculum of Van Helmont,	194
Spermatic cord,	219
fascia,	164, 169
Spheno-palatine ganglion,	64
Spinal arteries,	135
cord,	134, 136
nerves,	135
veins,	136
Spleen,	186
supplementary,	187
Spongy portion of the urethra,	217
Stapes,	96
Steno's duct,	24
Stomach,	185
Structure of bladder,	213
cornea,	76
intestinal tube,	184
kidney,	192
liver,	190
lungs,	125
œsophagus,	122
ovaries,	222
parotid gland,	24
prostate,	215
spleen,	186
stomach,	186
testicle,	218
tongue,	66
trachea,	71
uterus,	221
vagina,	220
vesiculæ seminales,	214
Striæ longitudinales,	86
Subarachnoid space,	80
Sublingual gland,	66
Submaxillary gland,	46
Substantia perforata,	87
Superficial fascia, of abdomen,	163, 168
of thigh,	195
Supra-renal capsules,	191
Suspensory ligament of liver,	188
of penis,	163
Symphysis pubes,	230
Synovial membrane of knee,	252
jaw,	49
Tænia semicircularis,	88
hippocampi,	89
Tapetum oculi,	77
Tarsal cartilages,	17
Tendo Achillis,	241
oculi,	17
Tendon, central, of diaphragm,	194
conjoined,	165, 166
Tentorium,	27, 28
Testes,	217
cerebri,	91
descent of,	257
Thalamus opticus,	88, 91
Thoracic duct,	55, 123
Throat, exploration of,	262
Thymus gland,	112
Thyro-hyoid membrane,	68
Thyroid axis,	58
cartilage,	70
body,	43
Tongue,	66
Tonsil,	63
Torcular Herophili,	27
Trabeculæ,	216
Trachea,	71
bifurcation of,	115
Tract, lateral,	84
optic,	82
Tragus,	14
Triangle of Hesselbach,	171
of Scarpa,	226
Triangles of the neck,	41
Triangular ligament,	204
Tricuspid valve,	118
Trigonum of the bladder,	213
Tube, Eustachian,	62
Tuber cinereum,	85
Tubuli galactophori,	101
seminiferi,	218
uriniferi,	192
Tunica albuginea,	218
vaginalis,	218
Tympanic bone,	95
Tympanum,	14, 94
Tyson's glands,	216
Umbilical region,	173
Urachus,	213
Ureter,	192
Urethra, male,	216
female,	220
Uterus,	221
Utricle,	69
Uvea,	77
Uvula of the bladder,	213
palate,	63
Vagina,	220
columns of the,	221
Vallecula,	92
Valsalva, sinuses of,	119
Valve, aortic,	119
arachnoid,	93
coronary,	117
Eustachian,	117, 238
ilio-cæcal,	183
mitral,	119
pyloric,	186
semilunar,	118, 119
sigmoid,	118
tricuspid,	118
venous,	223

INDEX

Valve—
  Vieussens, 92
Valvulæ conniventes, 184
Vasculum aberrans, 218
Vas deferens, 214, 219
Veins, axillary, 102
  azygos major, 123, 182, 195
    minor, 123, 182, 195
  basilic, 105
  cardiac, 116
  cava, inferior, 114, 182
    superior, 55, 113
  cephalic, 105
    internal, 51
  coronary, 116
  dorsalis penis, 215
  epigastric, 163
  facial, 23
  femoral, 226
  Galeni, 27
  gastric, 176
  hemorrhoidal, 210
  hepatic, 182
  iliac, 182
  innominata, 55, 112
  jugular, anterior, 39
    external, 38
    internal, 51, 55
  lumbar, 182
  maxillary, internal, 51
  median, 109
    basilic, 109
    cephalic, 109
  mesenteric, inferior, 176
    superior, 176
  ophthalmic, 29
  ovarian, 210
  phrenic, 182
  popliteal, 236
  portal, 176
  profunda, 228
  prostatic, 214
  pulmonary, 114
  radial, 109

Veins—
  renal, 182
  saphena, external, 241
    internal, 196, 223, 237
  spermatic, 181, 182, 210
  splenic, 176
  subclavian, 55
  supra-renal, 182
  Thebesii, 117
  thyroid, 55
  ulnar, 109
  umbilical, 257
  uterine, 221
  vertebral, 55
Velum interpositum, 89
  pendulum palati, 63
Venæ Galeni, 27, 90
Venesection, 109, 268
Ventricle of the brain, fifth, 89
  fourth, 93
  lateral, 87
  third, 90
Ventricles of the brain, 87
  heart, 117, 119
  of the larynx, 69
Vermiform processes, 92
Vertebral aponeurosis, 128
Veru montanum, 216
Vesiculæ seminales, 214
Vestibule, 97
Villi, 185
Vieussens, valve of, 92
Vincula subflava, 154
Vitreous humor, 179
Vulva, 219, 267

Wharton's duct, 47
Whartonian gelatine, 256
Willis, circle of, 81
Wilson's muscle, 211
Wrisberg, nerve of, 108

Zone of Zinn, 78
Zonula ciliaris, 78

CATALOGUE OF BOOKS

PUBLISHED BY

HENRY C. LEA.

(LATE LEA & BLANCHARD.)

The books in the annexed list will be sent by mail, post-paid, to any Post Office in the United States, on receipt of the printed prices. No risks of the mail, however, are assumed, either on money or books. Gentlemen will therefore, in most cases, find it more convenient to deal with the nearest bookseller.

Detailed catalogues furnished or sent free by mail on application. An illustrated catalogue of 64 octavo pages, handsomely printed, mailed on receipt of 10 cents. Address,

HENRY C. LEA,
Nos. 706 and 708 Sansom Street, Philadelphia.

PERIODICALS,

Free of Postage.

AMERICAN JOURNAL OF THE MEDICAL SCIENCES. Edited by Isaac Hays, M.D., published quarterly, about 1100 large 8vo. pages per annum,
MEDICAL NEWS AND LIBRARY, monthly, 384 large 8vo. pages per annum, } For five Dollars per annum in advance.

OR,

AMERICAN JOURNAL OF THE MEDICAL SCIENCES, Quarterly,
MEDICAL NEWS AND LIBRARY, monthly,
MONTHLY ABSTRACT OF MEDICAL SCIENCE, 48 pages per month, or nearly 600 pages per annum.
In all, about 2100 large 8vo. pages per annum, } For six Dollars per annum in advance.

MEDICAL NEWS AND LIBRARY, monthly, in advance, $1 00.

MONTHLY ABSTRACT OF MEDICAL SCIENCE, in advance, $2 50.

OBSTETRICAL JOURNAL. With an American Supplement, edited by WILLIAM F. JENKS, M.D. $5 00 per annum, in advance. Single Numbers, 50 cents. Is published monthly, each number containing about eighty octavo pages. Commencing with April, 1873.

ASHTON (T. J.) ON THE DISEASES, INJURIES, AND MALFORMATIONS OF THE RECTUM AND ANUS. With remarks on Habitual Constipation. Second American from the fourth London edition, with illustrations. 1 vol. 8vo. of about 300 pp. Cloth, $3 25.

ASHWELL (SAMUEL). A PRACTICAL TREATISE ON THE DISEASES OF WOMEN. Third American from the third London edition. In one 8vo. vol. of 528 pages. Cloth, $3 50.

ASHHURST (JOHN Jr.) THE PRINCIPLES AND PRACTICE OF SURGERY FOR THE USE OF STUDENTS AND PRACTITIONERS. In 1 large 8vo. vol. of over 1000 pages, containing 533 wood-cuts. Cloth, $6 50; leather, $7 50.

ATTFIELD (JOHN). CHEMISTRY, GENERAL, MEDICAL, AND PHARMACEUTICAL. Fifth edition, revised by the author. In 1 vol. 12mo. Cloth, $2 75; leather, $3 25.

ANDERSON (McCALL). ON THE TREATMENT OF DISEASES OF THE SKIN. In one small 8vo. vol. Cloth, $1 00.

BLOXAM (C. L.) CHEMISTRY, INORGANIC AND ORGANIC. With Experiments. In one handsome octavo volume of 700 pages, with 300 illustrations. Cloth, $4 00; leather, $5 00.

BRINTON (WILLIAM). LECTURES ON THE DISEASES OF THE STOMACH. From the second London edition, with illustrations. 1 vol. 8vo. Cloth, $3 25.

BRUNTON (T. LAUDER). A MANUAL OF MATERIA MEDICA AND THERAPEUTICS. In one 8vo. volume. (*Preparing*.)

BIGELOW (HENRY J.) ON DISLOCATION AND FRACTURE OF THE HIP, with the Reduction of the Dislocations by the Flexion Method. In one 8vo. vol. of 150 pp., with illustrations. Cloth, $2 50.

BASHAM (W. R.) RENAL DISEASES; A CLINICAL GUIDE TO THEIR DIAGNOSIS AND TREATMENT. With Illustrations. 1 vol. 12mo. Cloth, $2 00.

BUMSTEAD (F. J.) THE PATHOLOGY AND TREATMENT OF VENEREAL DISEASES. Including the results of recent investigations upon the subject. Third edition, revised and enlarged, with illustrations. 1 vol. 8vo., of over 700 pages. Cloth, $5; leather, $6.
—— AND CULLERIER'S ATLAS OF VENEREAL. See "CULLERIER."

BARLOW (GEORGE H.) A MANUAL OF THE PRACTICE OF MEDICINE. 1 vol. 8vo., of over 600 pages. Cloth, $2 50.

BAIRD (ROBERT). IMPRESSIONS AND EXPERIENCES OF THE WEST INDIES. 1 vol. royal 12mo. Cloth, 75 cents.

BARNES (ROBERT). A PRACTICAL TREATISE ON THE DISEASES OF WOMEN. In one handsome 8vo. vol. of about 800 pages, with 169 illustrations. Cloth, $5; leather, $6. (*Just issued*.)

BRYANT (THOMAS.) THE PRACTICE OF SURGERY. In one handsome octavo volume, of over 1000 pages, with many illustrations. Cloth, $6 25; leather, $7 25. (*Just issued*.)

BLANDFORD (G. FIELDING). INSANITY AND ITS TREATMENT. With an Appendix by Dr. Isaac Ray. In one handsome 8vo. vol., of 471 pages. Cloth, $3 25.

BELLAMY'S MANUAL OF SURGICAL ANATOMY. With numerous illustrations. In one royal 12mo. vol. Cloth, $2 25. (*Just issued*.)

BOWMAN (JOHN E.) A PRACTICAL HAND-BOOK OF MEDICAL CHEMISTRY. Edited by C. L. Bloxam. Sixth American, from the fourth and revised London edition. With numerous illustrations. 1 vol. royal 12mo. of 350 pages. Cloth, $2 25.

BOWMAN (JOHN E.) INTRODUCTION TO PRACTICAL CHEMISTRY, INCLUDING ANALYSIS. Edited by C. L. Bloxam. Sixth American, from the sixth and revised London edition, with numerous illustrations. 1 vol. royal 12mo. of 350 pages. Cloth, $2 25.

CHAMBERS (T. K.) THE INDIGESTIONS; OR, DISEASES OF THE DIGESTIVE ORGANS FUNCTIONALLY TREATED. Third American Edition, thoroughly revised by the author. 1 vol. 8vo., of over 300 pages. Cloth, $3 00.

—— RESTORATIVE MEDICINE. An Harveian Annual Oration delivered at the Royal College of Physicians, London, June 21, 1871. In one small 12mo. volume. Cloth, $1 00.

CARSON (JOSEPH). A SYNOPSIS OF THE COURSE OF LECTURES ON MATERIA MEDICA AND PHARMACY, delivered in the University of Pennsylvania. Fourth and revised edition. 1 vol. 8vo. Cloth, $3 00. (*Out of print for the present.*)

COOPER (B. B.) LECTURES ON THE PRINCIPLES AND PRACTICE OF SURGERY. In one large 8vo. vol. of 750 pages. Cloth, $2 00.

CARPENTER (WM. B.) PRINCIPLES OF HUMAN PHYSIOLOGY, WITH THEIR CHIEF APPLICATIONS TO PSYCHOLOGY, PATHOLOGY, THERAPEUTICS, HYGIENE, AND FORENSIC MEDICINE. In one large vol. 8vo., of nearly 900 closely printed pages. Cloth, $5 50; leather, raised bands, $6 50.

—— PRINCIPLES OF COMPARATIVE PHYSIOLOGY. New American, from the fourth and revised London edition. With over 300 beautiful illustrations. 1 vol. 8vo., of 752 pages. Cloth, $5.

—— PRIZE ESSAY ON THE USE OF ALCOHOLIC LIQUORS IN HEALTH AND DISEASE. New edition, with a Preface by D. F. Condie, M.D. 1 vol. 12mo. of 178 pages. Cloth, 60 cents.

CHRISTISON (ROBERT). DISPENSATORY OR COMMENTARY ON THE PHARMACOPŒIAS OF GREAT BRITAIN AND THE UNITED STATES. With a Supplement by R. E. Griffith. In one 8vo. vol. of over 1000 pages, containing 213 illustrations. Cloth, $4.

CHURCHILL (FLEETWOOD). ON THE THEORY AND PRACTICE OF MIDWIFERY. With notes and additions by D. Francis Condie, M.D. With about 200 illustrations. In one handsome 8vo. vol. of nearly 700 pages. Cloth, $4; leather, $5.

—— ESSAYS ON THE PUERPERAL FEVER, AND OTHER DISEASES PECULIAR TO WOMEN. In one neat octavo vol. of about 450 pages. Cloth, $2 50.

CONDIE (D. FRANCIS). A PRACTICAL TREATISE ON THE DISEASES OF CHILDREN. Sixth edition, revised and enlarged. In one large octavo volume of nearly 800 pages. Cloth, $5 25; leather, $6 25.

CULLERIER (A.) AN ATLAS OF VENEREAL DISEASES. Translated and edited by FREEMAN J. BUMSTEAD, M.D. A large imperial quarto volume, with 26 plates containing about 150 figures, beautifully colored, many of them the size of life. In one vol., strongly bound in cloth, $17.

—— Same work, in five parts, paper covers, for mailing, $3 per part.

CYCLOPEDIA OF PRACTICAL MEDICINE. By Dunglison, Forbes, Tweedie, and Conolly. In four large super royal octavo volumes, of 3254 double-columned pages, leather, raised bands, $15. Cloth, $11.

CAMPBELL'S LIVES OF LORDS KENYON, ELLENBOROUGH, AND TENTERDEN. Being the third volume of "Campbell's Lives of the Chief Justices of England." In one crown octavo vol. Cloth, $2.

DALTON (J. C.) A TREATISE ON HUMAN PHYSIOLOGY. Fifth edition, thoroughly revised, with 284 illustrations on wood. In one very handsome 8vo. vol. of over 700 pp. Cloth, $5 25; leather, $6 25.

DAVIS (F. H.) LECTURES ON CLINICAL MEDICINE. Second edition, revised and enlarged. In one 12mo. vol. Cloth, $1 75. (*Now ready.*)

DE JONGH, ON THE THREE KINDS OF COD-LIVER OIL. 1 small 12mo. vol., 75 cents.

DON QUIXOTE DE LA MANCHA. Illustrated edition. In two handsome vols. crown 8vo. Cloth, $2 50; half morocco, $3 70.

DEWEES (W. P.) A TREATISE ON THE DISEASES OF FEMALES. With illustrations. In one 8vo. vol. of 536 pages. Cloth, $3.

—— A TREATISE ON THE PHYSICAL AND MEDICAL TREATMENT OF CHILDREN. In one 8vo. vol. of 548 pages. Cloth, $2 80.

DRUITT (ROBERT). THE PRINCIPLES AND PRACTICE OF MODERN SURGERY. A revised American, from the eighth London edition. Illustrated with 432 wood engravings. In one 8vo. vol. of nearly 700 pages. Cloth, $4; leather, $5.

DUNGLISON (ROBLEY). MEDICAL LEXICON; a Dictionary of Medical Science. Containing a concise explanation of the various subjects and terms of Anatomy, Physiology, Pathology, Hygiene, Therapeutics, Pharmacology, Pharmacy, Surgery, Obstetrics, Medical Jurisprudence, and Dentistry. Notices of Climate and of Mineral Waters; Formulæ for Officinal, Empirical, and Dietetic Preparations, with the accentuation and Etymology of the Terms, and the French and other Synonymes. In one very large royal 8vo. vol. New edition. Cloth, $6 50; leather, $7 50. (*Just issued.*)

—— HUMAN PHYSIOLOGY. Eighth edition, thoroughly revised. In two large 8vo. vols. of about 1500 pp., with 532 illus. Cloth, $7.

—— NEW REMEDIES, WITH FORMULÆ FOR THEIR PREPARATION AND ADMINISTRATION. Seventh edition. In one very large 8vo. vol. of 770 pages. Cloth, $4.

DE LA BECHE'S GEOLOGICAL OBSERVER. In one large 8vo. vol. of 700 pages, with 300 illustrations. Cloth, $4.

DANA (JAMES D.) THE STRUCTURE AND CLASSIFICATION OF ZOOPHYTES. With illustrations on wood. In one imperial 4to. vol. Cloth, $4 00.

ELLIS (BENJAMIN). THE MEDICAL FORMULARY. Being a collection of prescriptions derived from the writings and practice of the most eminent physicians of America and Europe. Twelfth edition, carefully revised by A. H. Smith, M. D. In one 8vo. volume of 374 pages. Cloth, $3.

ERICHSEN (JOHN). THE SCIENCE AND ART OF SURGERY. A new and improved American, from the sixth enlarged and revised London edition. Illustrated with 630 engravings on wood. In two large 8vo. vols. Cloth, $9 00; leather, raised bands, $11 00.

ENCYCLOPÆDIA OF GEOGRAPHY. In three large 8vo. vols. Illustrated with 83 maps and about 1100 wood-cuts. Cloth, $5.

FENWICK (SAMUEL.) THE STUDENTS' GUIDE TO MEDICAL DIAGNOSIS. From the Third Revised and Enlarged London Edition. In one vol. royal 12mo., with numerous illustrations. Cloth, $2 25. (*Just issued.*)

FISKE FUND PRIZE ESSAYS ON TUBERCULOUS DISEASE. In one small 8vo. vol. Cloth, $1.

HENRY C. LEA'S PUBLICATIONS.

FLETCHER'S NOTES FROM NINEVEH, AND TRAVELS IN MESO-
POTAMIA, ASSYRIA, AND SYRIA. In one 12mo. vol. Cloth, 75 cts.

FOX ON DISEASES OF THE STOMACH. Publishing in the Medical News and Library for 1873 and 1874.

FLINT (AUSTIN). A TREATISE ON THE PRINCIPLES AND PRACTICE OF MEDICINE. Fourth edition, thoroughly revised and enlarged. In one large 8vo. volume of 1070 pages. Cloth, $6; leather, raised bands, $7. (*Just issued.*)

—— A PRACTICAL TREATISE, ON THE PHYSICAL EXPLORATION OF THE CHEST, AND THE DIAGNOSIS OF DISEASES AFFECTING THE RESPIRATORY ORGANS. Second and revised edition. One 8vo. vol. of 595 pages. Cloth, $4 50.

—— A PRACTICAL TREATISE ON THE DIAGNOSIS AND TREATMENT OF DISEASES OF THE HEART. Second edition, enlarged. In one neat 8vo. vol. of over 500 pages, $4 00.

—— MEDICAL ESSAYS. In one neat 12mo. volume. Cloth, $1 38. (*Now ready.*)

FOWNES (GEORGE). A MANUAL OF ELEMENTARY CHEMISTRY. From the tenth enlarged English edition. In one royal 12mo. vol. of 857 pages, with 197 illustrations. Cloth, $2 75; leather, $3 25.

FULLER (HENRY). ON DISEASES OF THE LUNGS AND AIR PASSAGES. Their Pathology, Physical Diagnosis, Symptoms and Treatment. From the second English edition. In one 8vo. vol. of about 500 pages. Cloth, $3 50.

GALLOWAY (ROBERT). A MANUAL OF QUALITATIVE ANALYSIS. From the fifth English edition. In one 12mo. vol. Cloth, $2 50. (*Lately published.*)

GLUGE (GOTTLIEB). ATLAS OF PATHOLOGICAL HISTOLOGY. Translated by Joseph Leidy, M.D., Professor of Anatomy in the University of Pennsylvania, &c. In one vol. imperial quarto, with 320 copper plate figures, plain and colored. Cloth, $4.

GREEN (T. HENRY). AN INTRODUCTION TO PATHOLOGY AND MORBID ANATOMY. In one handsome 8vo. vol., with numerous illustrations. Cloth, $2 50.

GIBSON'S INSTITUTES AND PRACTICE OF SURGERY. In two 8vo. vols. of about 1000 pages, leather, $6 50.

GRAY (HENRY). ANATOMY, DESCRIPTIVE AND SURGICAL. A new American, from the fifth and enlarged London edition. In one large imperial 8vo. vol. of about 900 pages, with 462 large and elaborate engravings on wood. Cloth, $6; leather, $7. (*Just issued.*)

GRIFFITH (ROBERT E.) A UNIVERSAL FORMULARY, CONTAINING THE METHODS OF PREPARING AND ADMINISTERING OFFICINAL AND OTHER MEDICINES. Third and Enlarged edition. Edited by John M. Maisch. In one large 8vo. vol. of 800 pages, double columns. Cloth, $4 50; leather, $5 50.

GROSS (SAMUEL D.) A SYSTEM OF SURGERY, PATHOLOGICAL, DIAGNOSTIC, THERAPEUTIC, AND OPERATIVE. Illustrated by 1403 engravings. Fifth edition, revised and improved. In two large imperial 8vo. vols. of over 2200 pages, strongly bound in leather, raised bands. $15. (*Lately issued.*)

—— A PRACTICAL TREATISE ON FOREIGN BODIES IN THE AIR PASSAGES. In one 8vo. vol. of 468 pages. Cloth, $2 75.

—— ELEMENTS OF PATHOLOGICAL ANATOMY. Third edition. In one large 8vo. vol. of nearly 800 pages, with about 350 illustrations. Cloth, $4.

GUERSANT (P.) SURGICAL DISEASES OF INFANTS AND CHILDREN. Translated by R. J. Dunglison, M. D. In one 8vo. vol. Cloth, $2 50.

HUDSON (A.) LECTURES ON THE STUDY OF FEVER. 1 vol. 8vo., 316 pages. Cloth, $2 50.

HEATH (CHRISTOPHER). PRACTICAL ANATOMY; A MANUAL OF DISSECTIONS. With additions, by W. W. Keen, M. D. In 1 volume; with 247 illustrations. Cloth, $3 50; leather, $4.

HARTSHORNE (HENRY). ESSENTIALS OF THE PRINCIPLES AND PRACTICE OF MEDICINE. Fourth and revised edition. In one 12mo. vol. Cloth, $2 63; half bound, $2 88. (*Now ready*.)

——— CONSPECTUS OF THE MEDICAL SCIENCES. Comprising Manuals of Anatomy, Physiology, Chemistry, Materia Medica, Practice of Medicine, Surgery, and Obstetrics. Second Edition. In one royal 12mo. volume of over 1000 pages, with 477 illustrations. Strongly bound in leather, $5 00; cloth, $4 25. (*Now ready*.)

——— A HANDBOOK OF ANATOMY AND PHYSIOLOGY. In one neat royal 12mo. volume, with many illustrations. Cloth, $1 75. (*Just ready*.)

HAMILTON (FRANK H.) A PRACTICAL TREATISE ON FRACTURES AND DISLOCATIONS. Fourth and revised edition. In one handsome 8vo. vol. of 789 pages, and 322 illustrations. Cloth, $5 75; leather, $6 75.

HOLMES (TIMOTHY). A MANUAL OF PRACTICAL SURGERY. In one handsome 8vo. volume, with many illustrations. (*Preparing*.)

HOBLYN (RICHARD D.) A DICTIONARY OF THE TERMS USED IN MEDICINE AND THE COLLATERAL SCIENCES. In one 12mo. volume, of over 500 double columned pages. Cloth, $1 50; leather, $2.

HODGE (HUGH L.) ON DISEASES PECULIAR TO WOMEN, INCLUDING DISPLACEMENTS OF THE UTERUS. Second and revised edition. In one 8vo. volume. Cloth, $4 50.

——— THE PRINCIPLES AND PRACTICE OF OBSTETRICS. Illustrated with large lithographic plates containing 159 figures from original photographs, and with numerous wood-cuts. In one large quarto vol. of 550 double-columned pages. Strongly bound in cloth, $14.

HOLLAND (SIR HENRY). MEDICAL NOTES AND REFLECTIONS. From the third English edition. In one 8vo. vol. of about 500 pages. Cloth, $3 50.

HODGES (RICHARD M.) PRACTICAL DISSECTIONS. Second edition. In one neat royal 12mo. vol., half bound, $2.

HUGHES SCRIPTURE GEOGRAPHY AND HISTORY, with 12 colored maps. In 1 vol. 12mo. Cloth, $1.

HORNER (WILLIAM E.) SPECIAL ANATOMY AND HISTOLOGY. Eighth edition, revised and modified. In two large 8vo. vols. of over 1000 pages, containing 300 wood-cuts. Cloth, $6.

HILL (BERKELEY). SYPHILIS AND LOCAL CONTAGIOUS DISORDERS. In one 8vo. volume of 467 pages. Cloth, $3 25.

HILLIER (THOMAS). HAND-BOOK OF SKIN DISEASES. Second Edition. In one neat royal 12mo. volume, about 300 pp., with two plates. Cloth, $2 25.

HALL (MRS. M.) LIVES OF THE QUEENS OF ENGLAND BEFORE THE NORMAN CONQUEST. In one handsome 8vo. vol. Cloth, $2 25; crimson cloth, $2 50; half morocco, $3.

JONES (C. HANDFIELD), AND SIEVEKING (E. D. H.) A MANUAL OF PATHOLOGICAL ANATOMY. In one large 8vo. vol. of nearly 750 pages, with 397 illustrations. Cloth, $3 50.

JONES (C. HANDFIELD). CLINICAL OBSERVATIONS ON FUNCTIONAL NERVOUS DISORDERS. Second American Edition. In one 8vo. vol. of 348 pages. Cloth, $3 25.

KIRKES (WILLIAM SENHOUSE). A MANUAL OF PHYSIOLOGY. A new American, from the eighth London edition. One vol., with many illus., 12mo. Cloth, $3 25; leather, $3 75. (*Lately issued.*)

KNAPP (F.) TECHNOLOGY; OR CHEMISTRY, APPLIED TO THE ARTS AND TO MANUFACTURES, with American additions, by Prof. Walter R. Johnson. In two 8vo. vols., with 500 ill. Cloth, $6.

KENNEDY'S MEMOIRS OF THE LIFE OF WILLIAM WIRT. In two vols. 12mo. Cloth, $2.

LEA (HENRY C.) SUPERSTITION AND FORCE; ESSAYS ON THE WAGER OF LAW, THE WAGER OF BATTLE, THE ORDEAL, AND TORTURE. Second edition, revised. In one handsome royal 12mo. vol., $2 75.

—— STUDIES IN CHURCH HISTORY. The Rise of the Temporal Power—Benefit of Clergy—Excommunication. In one handsome 12mo. vol. of 515 pp. Cloth, $2 75.

—— AN HISTORICAL SKETCH OF SACERDOTAL CELIBACY IN THE CHRISTIAN CHURCH. In one handsome octavo volume of 602 pages. Cloth, $3 75.

LA ROCHE (R.) YELLOW FEVER IN ITS HISTORICAL, PATHOLOGICAL, ETIOLOGICAL, AND THERAPEUTICAL RELATIONS. In two 8vo. vols. of nearly 1500 pages. Cloth, $7.

—— PNEUMONIA, ITS SUPPOSED CONNECTION, PATHOLOGICAL AND ETIOLOGICAL, WITH AUTUMNAL FEVERS. In one 8vo. vol. of 500 pages. Cloth, $3.

LINCOLN (D. F.) ELECTRO-THERAPEUTICS. A Condensed Manual of Medical Electricity. In one neat royal 12mo. volume, with illustrations. Cloth, $1 50. (*Just issued.*)

LEISHMAN (WILLIAM). A SYSTEM OF MIDWIFERY. Including the Diseases of Pregnancy and the Puerperal State. In one large and very handsome 8vo. vol. of 700 pages and 182 illus. Cloth, $5; leather, $6.

LAURENCE (J. Z.) AND MOON (ROBERT C.) A HANDY-BOOK OF OPHTHALMIC SURGERY. Second edition, revised by Mr. Laurence. With numerous illus. In one 8vo. vol. Cloth, $2 75.

LEHMANN (C. G.) PHYSIOLOGICAL CHEMISTRY. Translated by George F. Day, M. D. With plates, and nearly 200 illustrations. In two large 8vo. vols., containing 1200 pages. Cloth, $6.

—— A MANUAL OF CHEMICAL PHYSIOLOGY. In one very handsome 8vo. vol. of 356 pages. Cloth, $2 25.

LAWSON (GEORGE). INJURIES OF THE EYE, ORBIT, AND EYELIDS, with about 100 illustrations. From the last English edition. In one handsome 8vo. vol. Cloth, $3 50.

LUDLOW (J. L.) A MANUAL OF EXAMINATIONS UPON ANATOMY, PHYSIOLOGY, SURGERY, PRACTICE OF MEDICINE, OBSTETRICS, MATERIA MEDICA, CHEMISTRY, PHARMACY, AND THERAPEUTICS. To which is added a Medical Formulary. Third edition. In one royal 12mo. vol. of over 800 pages. Cloth, $3 25; leather, $3 75.

LYNCH (W. F.) A NARRATIVE OF THE UNITED STATES EX-
PEDITION TO THE DEAD SEA AND RIVER JORDAN. In one
large and handsome octavo vol., with 28 beautiful plates and two
maps. Cloth, $3.
—— Same Work, condensed edition. One vol. royal 12mo. Cloth, $1.

LAYCOCK (THOMAS). LECTURES ON THE PRINCIPLES AND
METHODS OF MEDICAL OBSERVATION AND RESEARCH. In
one 12mo. vol. Cloth, $1.

LYONS (ROBERT D.) A TREATISE ON FEVER. In one neat 8vo.
vol. of 362 pages. Cloth, $2 25.

MARSHALL (JOHN). OUTLINES OF PHYSIOLOGY, HUMAN
AND COMPARATIVE. With Additions by FRANCIS G. SMITH,
M. D., Professor of the Institutes of Medicine in the University of
Pennsylvania. In one 8vo. volume of 1026 pages, with 122 illustra-
tions. Strongly bound in leather, raised bands, $7 50. Cloth, $6 50.

MACLISE (JOSEPH). SURGICAL ANATOMY. In one large imperial quarto vol., with 68 splendid plates, beautifully colored; containing 190 figures, many of them life size. Cloth, $14.

MEIGS (CHAS. D). WOMAN: HER DISEASES AND THEIR REM-
EDIES. Fourth and improved edition. In one large 8vo. vol. of
over 700 pages. Cloth, $5; leather, $6.

—— ON THE NATURE, SIGNS, AND TREATMENT OF CHILD-BED
FEVER In one 8vo. vol. of 365 pages. Cloth, $2.

MILLER (JAMES). PRINCIPLES OF SURGERY. Fourth American,
from the third Edinburgh edition. In one large 8vo. vol. of 700
pages, with 240 illustrations. Cloth, $3 75.

—— THE PRACTICE OF SURGERY. Fourth American, from the
last Edinburgh edition. In one large 8vo. vol. of 700 pages, with
364 illustrations. Cloth, $3 75.

MONTGOMERY (W. F.) AN EXPOSITION OF THE SIGNS AND
SYMPTOMS OF PREGNANCY. From the second English edition.
In one handsome 8vo. vol. of nearly 600 pages. Cloth, $3 75.

MULLER (J.) PRINCIPLES OF PHYSICS AND METEOROLOGY.
In one large 8vo. vol. with 550 wood-cuts, and two colored plates.
Cloth, $4 50.

MIRABEAU; A LIFE HISTORY. In one royal 12mo. vol. Cloth,
75 cents.

MACFARLAND'S TURKEY AND ITS DESTINY. In 2 vols. royal
12mo. Cloth, $2.

MARSH (MRS.) A HISTORY OF THE PROTESTANT REFORMA-
TION IN FRANCE. In 2 vols. royal 12mo. Cloth, $2.

NELIGAN (J. MOORE). A PRACTICAL TREATISE ON DISEASES
OF THE SKIN. Fifth American, from the second Dublin edition.
In one neat royal 12mo. vol. of 462 pages. Cloth, $2 25. (*Out
of print for the present.*)

—— AN ATLAS OF CUTANEOUS DISEASES. In one handsome
quarto vol. with beautifully colored plates, &c. Cloth, $5 50.

NEILL (JOHN) AND SMITH (FRANCIS G.) COMPENDIUM OF
THE VARIOUS BRANCHES OF MEDICAL SCIENCE. In one
handsome 12mo. vol. of about 1000 pages, with 374 wood-cuts.
Cloth, $4; leather, raised bands, $4 75.

NIEBUHR (B. G.) LECTURES ON ANCIENT HISTORY; comprising the history of the Asiatic Nations, the Egyptians, Greeks, Macedonians, and Carthagenians. Translated by Dr. L. Schmitz. In three neat volumes, crown octavo. Cloth, $5 00.

ODLING (WILLIAM). A COURSE OF PRACTICAL CHEMISTRY FOR THE USE OF MEDICAL STUDENTS. From the fourth revised London edition. In one 12mo. vol. of 261 pp., with 75 illustrations. Cloth, $2.

PAVY (F. W.) A TREATISE ON THE FUNCTION OF DIGESTION: ITS DISORDERS AND THEIR TREATMENT. From the second London Ed. In one 8vo. vol. of 246 pp. Cloth, $2. (*Lately Issued.*)

——— A TREATISE ON FOOD AND DIETETICS PHYSIOLOGICALLY AND THERAPEUTICALLY CONSIDERED. In one neat octavo volume of about 500 pages. Cloth, $4 75. (*Just issued.*)

PARRISH (EDWARD). A TREATISE ON PHARMACY. With many Formulæ and Prescriptions. Fourth edition. Enlarged and thoroughly revised by Thomas S. Wiegand. In one handsome 8vo. vol. of 977 pages, with 280 illus. Cloth, $5 50; leather, $6 50. (*Just issued.*)

PIRRIE (WILLIAM) THE PRINCIPLES AND PRACTICE OF SURGERY. In one handsome octavo volume of 780 pages, with 316 illustrations. Cloth, $3 75.

PEREIRA (JONATHAN). MATERIA MEDICA AND THERAPEUTICS. An abridged edition. With numerous additions and references to the United States Pharmacopœia. By Horatio C. Wood, M. D. In one large octavo volume, of 1040 pages, with 236 illustrations. Cloth $7 00; leather, raised bands, $8 00.

PULSZKY'S MEMOIRS OF AN HUNGARIAN LADY. In one neat royal 12mo. vol. Cloth, $1.

PAGET'S HUNGARY AND TRANSYLVANIA. In two royal 12mo. vols. Cloth, $2.

ROBERTS (WILLIAM). A PRACTICAL TREATISE ON URINARY AND RENAL DISEASES. A second American, from the second London edition. With numerous illustrations and a colored plate. In one very handsome 8vo. vol. of 616 pages. Cloth, $4 50. (*Just Issued.*)

RAMSBOTHAM (FRANCIS H.) THE PRINCIPLES AND PRACTICE OF OBSTETRIC MEDICINE AND SURGERY. In one imperial 8vo. vol. of 650 pages, with 64 plates, besides numerous woodcuts in the text. Strongly bound in leather $7.

RIGBY (EDWARD). A SYSTEM OF MIDWIFERY. Second American edition. In one handsome 8vo. vol. of 422 pages. Cloth, $2 50.

RANKE'S HISTORY OF THE TURKISH AND SPANISH EMPIRES in the 16th and beginning of 17th Century. In one 8vo. volume, paper, 25 cts.

——— HISTORY OF THE REFORMATION IN GERMANY. Parts I. II. III. In one vol. Cloth, $1.

ROYLE (J. FORBES). MATERIA MEDICA AND THERAPEUTICS. Edited by Jos. Carson, M. D. In one large 8vo. vol. of about 700 pages, with 98 illustrations. Cloth, $3.

SMITH (EUSTACE). ON THE WASTING DISEASES OF CHILDREN. Second American edition, enlarged. In one 8vo. vol. Cloth, $2 50. (*Just Issued.*)

SMITH (J. LEWIS.) A TREATISE ON THE DISEASES OF IN-
FANCY AND CHILDHOOD. Second edition. In one large 8vo.
volume of over 700 pages. Cloth, $5; leather, $6.

SARGENT (F. W.) ON BANDAGING AND OTHER OPERATIONS
OF MINOR SURGERY. New edition, with an additional chapter
on Military Surgery. In one handsome royal 12mo. vol. of nearly
400 pages, with 184 wood-cuts. Cloth, $1 75.

SHARPEY (WILLIAM) AND QUAIN (JONES AND RICHARD).
HUMAN ANATOMY. With notes and additions by Jos. Leidy,
M. D., Prof. of Anatomy in the University of Pennsylvania. In two
large 8vo. vols. of about 1300 pages, with 511 illustrations. Coth, $6.

SKEY (FREDERIC C.) OPERATIVE SURGERY. In one 8vo. vol.
of over 650 pages, with about 100 wood-cuts. Cloth, $3 25.

SLADE (D. D.) DIPHTHERIA; ITS NATURE AND TREATMENT.
Second edition. In one neat royal 12mo. vol. Cloth, $1 25.

SMITH (HENRY H.) AND HORNER (WILLIAM E.) ANATOMICAL
ATLAS. Illustrative of the structure of the Human Body. In one large
imperial 8vo. vol., with about 650 beautiful figures. Cloth, $4 50.

SMITH (EDWARD). CONSUMPTION; ITS EARLY AND REME-
DIABLE STAGES. In one 8vo. vol. of 254 pp. Cloth, $2 25.

STILLE (ALFRED). THERAPEUTICS AND MATERIA MEDICA.
Fourth edition, revised and enlarged. In two large and handsome
volumes 8vo. Cloth, $10; leather, $12. (*Now ready.*)

SCHMITZ AND ZUMPT'S CLASSICAL SERIES. In royal 18mo.
CORNELII NEPOTIS LIBER DE EXCELLENTIBUS DUCIBUS
EXTERARUM GENTIUM, CUM VITIS CATONIS ET ATTICI.
With notes, &c. Price in cloth, 60 cents; half bound, 70 cts.

C. I. CÆSARIS COMMENTARII DE BELLO GALLICO. With notes,
map, and other illustrations. Price in cloth, 60 cents; half bound,
70 cents.

C. C. SALLUSTII DE BELLO CATILINARIO ET JUGURTHINO.
With notes, map, &c. Price in cloth, 60 cents; half bound, 70 cents.

Q. CURTII RUFII DE GESTIS ALEXANDRI MAGNI LIBRI VIII.
With notes, map, &c. Price in cloth, 80 cents; half bound, 90 cents.

P. VIRGILII MARONIS CARMINA OMNIA. Price in cloth, 85
cents; half bound, $1.

M. T. CICERONIS ORATIONES SELECTÆ XII. With notes, &c.
Price in cloth, 70 cents; half bound, 80 cents.

ECLOGÆ EX Q. HORATII FLACCI POEMATIBUS. With notes,
&c. Price in cloth, 70 cents; half bound, 80 cents.

ADVANCED LATIN EXERCISES, WITH SELECTIONS FOR
READING. Revised, with additions. Cloth, price 60 cents; half
bound, 70 cents.

SWAYNE (JOSEPH GRIFFITHS). OBSTETRIC APHORISMS. A
new American, from the fifth revised English edition. With additions by E. R. Hutchins, M. D. In one small 12mo. vol. of 177 pp.,
with illustrations. Cloth, $1 25.

STURGES (OCTAVIUS). AN INTRODUCTION TO THE STUDY
OF CLINICAL MEDICINE. In one 12mo. vol. Cloth, $1 25.
(*Lately published.*)

SCHOEDLER (FREDERICK) AND MEDLOCK (HENRY). WONDERS OF NATURE. An elementary introduction to the Sciences of Physics, Astronomy, Chemistry, Mineralogy, Geology, Botany, Zoology, and Physiology. Translated from the German by H. Medlock. In one neat 8vo. vol., with 679 illustrations. Cloth, $3.

STOKES (W.) LECTURES ON FEVER. One handsome 8vo. volume. (*Preparing.*)

SMALL BOOKS ON GREAT SUBJECTS. Twelve works; each one 10 cents, sewed, forming a neat and cheap series; or done up in 3 vols. Cloth, $1 50.

STRICKLAND (AGNES). LIVES OF THE QUEENS OF HENRY THE VIII. AND OF HIS MOTHER. In one crown octavo vol., extra cloth, $1; black cloth, 90 cents.

—— MEMOIRS OF ELIZABETH, SECOND QUEEN REGNANT OF ENGLAND AND IRELAND. In one crown octavo vol., extra cloth, $1 40; black cloth, $1 30.

TANNER (THOMAS HAWKES). A MANUAL OF CLINICAL MEDICINE AND PHYSICAL DIAGNOSIS. Third American from the second revised English edition. Edited by Tilbury Fox, M.D. In one handsome 12mo. volume of 366 pp. Cloth, $1 50.

—— ON THE SIGNS AND DISEASES OF PREGNANCY. First American from the second English edition. With four colored plates and numerous illustrations on wood. In one vol. 8vo. of about 500 pages. Cloth, $4 25.

TUKE (DANIEL HACK). INFLUENCE OF THE MIND UPON THE BODY. In one handsome 8vo. vol. of 416 pp. Cloth, $3 25. (*Just issued.*)

TAYLOR (ALFRED S.) MEDICAL JURISPRUDENCE. Seventh American edition. Edited by John J. Reese, M.D. In one large 8vo. volume of 879 pages. Cloth, $5; leather, $6. (*Just issued.*)

—— PRINCIPLES AND PRACTICE OF MEDICAL JURISPRUDENCE. From the Second English Edition. In two large 8vo. vols. Cloth, $10; leather, $12. (*Just issued.*)

—— ON POISONS IN RELATION TO MEDICINE AND MEDICAL JURISPRUDENCE. Third American from the Third London Edition. 1 vol. 8vo. (*Preparing.*)

THOMAS (T. GAILLARD). A PRACTICAL TREATISE ON THE DISEASES OF FEMALES. Fourth edition, thoroughly revised. In one large and handsome octavo volume of 801 pages, with 191 illustrations. Cloth, $5 00; leather, $6 00. (*Now ready.*)

TODD (ROBERT BENTLEY). CLINICAL LECTURES ON CERTAIN ACUTE DISEASES. In one vol. 8vo. of 320 pp., extra cloth, $2 50.

THOMPSON (SIR HENRY). CLINICAL LECTURES ON DISEASES OF THE URINARY ORGANS. Second and revised edition. In one 8vo. volume, with illustrations. Cloth, $2 25. (*Now ready.*)

—— THE PATHOLOGY AND TREATMENT OF STRICTURE OF THE URETHRA AND URINARY FISTULÆ. From the third English edition. In one 8vo. vol. of 359 pp., with illus. Cloth, $3 50.

—— THE DISEASES OF THE PROSTATE, THEIR PATHOLOGY AND TREATMENT. Fourth edition, revised. In one very handsome 8vo. vol. of 355 pp., with 13 plates. Cloth, $3 75.

WALSHE (W. H.) PRACTICAL TREATISE ON THE DISEASES OF THE HEART AND GREAT VESSELS. Third American from the third revised London edition. In one 8vo. vol. of 420 pages. Cloth, $3.

WOHLER'S OUTLINES OF ORGANIC CHEMISTRY. Translated from the 8th German edition, by Ira Remsen, M.D. In one neat 12mo. vol. Cloth, $3 00. (*Lately issued.*)

WALES (PHILIP S.) MECHANICAL THERAPEUTICS. In one large 8vo. vol. of about 700 pages, with 642 illustrations on wood. Cloth, $5 75; leather, $6 75.

WELLS (J. SOELBERG). A TREATISE ON THE DISEASES OF THE EYE. Second American, from the Third English edition, with additions by I. Minis Hays, M.D. In one large and handsome octavo vol., with 6 colored plates and many wood-cuts, also selections from the test-types of Jaeger and Snellen. Cloth, $5 00; leather, $6 00. (*Lately issued.*)

WHAT TO OBSERVE AT THE BEDSIDE AND AFTER DEATH IN MEDICAL CASES. In one royal 12mo. vol. Cloth, $1.

WATSON (THOMAS). LECTURES ON THE PRINCIPLES AND PRACTICE OF PHYSIC. A new American from the fifth and enlarged English edition, with additions by H. Hartshorne, M.D. In two large and handsome octavo volumes. Cloth, $9; leather, $11. (*Lately issued.*)

WEST (CHARLES). LECTURES ON THE DISEASES PECULIAR TO WOMEN. Third American from the Third English edition. In one octavo volume of 550 pages. Cloth, $3 75; leather, $4 75.

—— **LECTURES ON THE DISEASES OF INFANCY AND CHILDHOOD.** Fifth American from the sixth revised English edition. In one large 8vo. vol. of 670 closely printed pages. Cloth, $4 50; leather, $5 50. (*Now ready.*)

—— **ON SOME DISORDERS OF THE NERVOUS SYSTEM IN CHILDHOOD.** From the London Edition. In one small 12mo. volume. Cloth, $1.

—— **AN ENQUIRY INTO THE PATHOLOGICAL IMPORTANCE OF ULCERATION OF THE OS UTERI.** In one vol. 8vo. Cloth, $1 25.

WILLIAMS (CHARLES J. B.) PULMONARY CONSUMPTION: ITS NATURE, VARIETIES, AND TREATMENT. In one neat octavo volume. Cloth, $2 50. (*Lately publised.*)

WILSON (ERASMUS). A SYSTEM OF HUMAN ANATOMY. A new and revised American from the last English edition. Illustrated with 397 engravings on wood. In one handsome 8vo. vol. of over 600 pages. Cloth, $4; leather, $5.

—— **ON DISEASES OF THE SKIN.** The seventh American from the last English edition. In one large 8vo. vol. of over 800 pages. Cloth, $5.

Also, A SERIES OF PLATES, illustrating "Wilson on Diseases of the Skin," consisting of 20 plates, thirteen of which are beautifully colored, representing about one hundred varieties of Disease. $5 50.

Also, the TEXT AND PLATES, bound in one volume. Cloth, $10.

—— **THE STUDENT'S BOOK OF CUTANEOUS MEDICINE.** In one handsome royal 12mo. vol. Cloth, $3 50.

WINSLOW (FORBES). ON OBSCURE DISEASES OF THE BRAIN AND DISORDERS OF THE MIND. In one handsome 8vo. vol. of nearly 600 pages. Cloth, $4 25.

WINCKEL ON DISEASES OF CHILDBED. Translated by Chadwick. (*Preparing.*)

ZEISSL ON VENEREAL DISEASES. Translated by Sturgis. (*Preparing.*)

www.ingramcontent.com/pod-product-compliance
Lightning Source LLC
Chambersburg PA
CBHW030235240426
43663CB00037B/853